I0063311

Strategies for Higher Thinking

Optimizing Our Lives by Embracing AI

DWAINE AJ WHOGOES

DIRECTLIVING
PUBLISHING COMPANY
Sustainable Growth Through Creative
GROWTH THROUGH CREATIVE PROCESS

DIRECTLIVING PUBLISHING COMPANY 2024

Copyright © 2024 by Dwaine AJ Whogoes. All rights reserved.

This publication is protected by copyright law. It's crucial to respect intellectual property. No part of this publication may be reproduced, distributed, or transmitted in any form or by any means, including photocopying, recording, or other electronic or mechanical methods, without the publisher's prior written permission, except as permitted by U.S. copyright law. For permission requests, contact Directliving Publishing Company / Dwaine AJ Whogoes.

This publication is a work of fiction. The story, all names, characters, and incidents portrayed are purely imaginative, and any resemblance to actual persons (living or deceased), places, buildings, and products is strictly coincidental and unintended.

Book Cover by cemerbas
Illustrations by Dwaine AJ Whogoes
First edition 2024

ISBN - 9798332265433 - Paperback
ISBN - 9798332424472 - Hardcover

Literary Titan Award Winner

Barnes & Noble - Review
Literary Titan Review
☆☆☆☆☆
5 out of 5 stars
Explaining AI In Way Everyone Can Understand

"The topic of artificial intelligence (AI) is undeniably divisive, with some opposing it due to ethical concerns and its potential to displace jobs, while others champion its ability to significantly enhance our lives. In Strategies for Higher Thinking, Dwaine AJ Whogoes presents his thoughtful perspective on AI, exploring its profound impact on our daily lives. His approach to the subject is both enlightening and accessible, carefully avoiding any conde-scension toward those with differing views. Instead, Whogoes aims to educate a broad audience about a technology that is clearly here to stay.

The book begins by highlighting how deeply embedded AI is in our everyday activities, from using smart devices to browsing online, effectively setting the stage for a more in-depth exploration of AI's revolutionary impact across various sectors. Whogoes offers fascinating insights into the practical applications of AI, particularly in everyday devices like smartphones. He explains how AI enhances functionalities such as facial recognition, predictive text, camera features, and app personalization. Additionally, the author delves into AI's transformative role in smart homes, healthcare, and social media, emphasizing its potential to improve efficiency, security, and personalization in our daily lives.

Whogoes' enthusiasm for the subject is infectious, keeping readers engaged throughout the book. His authority and passion for AI are evident as he demonstrates its potential to solve complex problems and significantly improve our lives. His primary goal is to make AI accessible and understandable to everyone, regardless of their background. The book successfully demystifies AI through practical examples, ethical discussions, and interactive activities, making even the most complex concepts comprehensible and engaging. For those who may have previously avoided discussing AI due to a lack of understanding, this book provides a solid foundation for forming informed opinions and choosing a stance on the subject. Whogoes' writing is inclusive, targeting a diverse audience that includes teachers, entrepreneurs, technologists, and job seekers. He adheres to the basics and facts about AI, with each chapter building on the last to ensure a comprehensive understanding of AI's capabilities and its future role. While one might expect the book to lean heavily toward advocacy, Whogoes remains neutral and focused on informing rather than persuading, allowing readers to draw their own conclusions.

Strategies for Higher Thinking is a highly informative read that breaks down what can often be perceived as a complex subject into simple, digestible parts that anyone can understand, even those unfamiliar with AI. This book is a valuable addition to the growing body of literature on how technology can drive positive change and inspire innovative thinking. Given the increasing integration of AI into everyday life, Whogoes' work is a timely resource that challenges readers to rethink their positions or, at the very least, support their arguments with well-reasoned facts. "

☆☆☆☆☆
5 out of 5 stars.
Literary Titan Award Winner

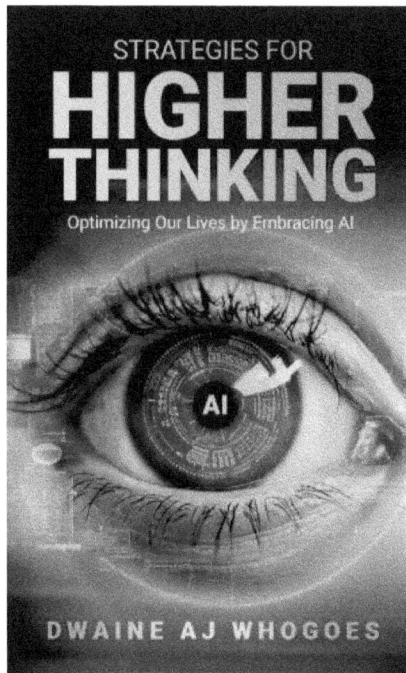

Strategies For Higher Thinking: Optimizing Our Lives by Embracing AI
by Dwaine Aj Whogoes

Preface

Artificial Intelligence
a public enigma for the modern age
scientifically fascinating, engineering bewildering
a vastness of possibilities and ethical complexities
rooted in early computer science and cognitive psychology
we have entered a new age of optimizing opportunities
and unlimited capabilities in human potential

D AJ Whogoes

Table of Contents

Introduction

Every day, whether browsing online, snapping photos to share with friends, or just asking your smart device for the weather, you're interacting with artificial intelligence (AI). But what if I told you these interactions are just the tip of the iceberg? Imagine AI simplifying your daily tasks and revolutionizing how you work, enhancing educational methods, or even predicting health issues before they become apparent. This is the transformative journey of AI that we are on—a journey that has the potential to reshape every facet of our lives.

My enthusiasm for AI is not just about the technology itself but its vast potential to solve complex problems and significantly improve our lives. As someone deeply immersed in AI, I've witnessed firsthand how this technology can drive innovation and offer solutions that were once thought impossible. In writing this book, I aim to open up this world to you, making it accessible and understandable, regardless of your profession or background.

Strategies for Higher Thinking: Optimizing Our Lives by Embracing AI is designed to demystify AI for a broad audience. It provides practical examples of AI in action, addresses ethical considerations, and encourages the responsible development and use of AI technologies. This book is unique because it simplifies complex AI concepts and addresses common myths to help solidify your understanding of AI.

This book is for teachers eager to integrate AI into their classrooms, entrepreneurs looking to leverage AI for business innovations, technologists developing AI solutions, and job seekers curious about AI's impact on the job market. Essentially, it is for anyone interested in understanding and engaging with this pivotal technology.

The book is structured to guide you through the basics to more advanced implications of AI. It explores practical applications, ethical considerations, and future trends. Each chapter builds on the previous one, ensuring a comprehensive understanding of AI's capabilities and role in our future.

Engaging with AI is about understanding technology and empowering yourself to think differently, innovate, and solve real-world problems. AI is a tool for positive change and advancement, and by embracing it, you can make a significant impact in your field and beyond.

It is natural to feel apprehensive about new technologies. Throughout this book, I address these fears and uncertainties by providing clear, factual information and balanced perspectives. By sharing personal anecdotes and insights from my journey with AI, I aim to make the narrative informative and relatable.

So, I invite you to open your mind to the possibilities of AI. With *Strategies for Higher Thinking: Optimizing Our Lives by Embracing AI* as your guide, you can navigate the future confidently and responsibly, ready to embrace AI's myriad opportunities. Let's embark on this exciting journey together, exploring how AI can enhance our lives and inspire us to rethink what is possible.

ONE

Demystifying AI for Everyday Use

D id you know the device in your pocket is a bustling artificial intelligence hub, quietly working behind the scenes to simplify and enhance your daily activities? Most of us interact with AI far more often than we realize. From unlocking our phones with a glance to receiving movie recommendations, AI's invisible hand is omnipresent. This chapter will peel back the curtain on the AI functionalities embedded in your smartphone, revealing how these advanced technologies shape your interaction with the digital world, often without your conscious awareness. By understanding these mechanisms, you can better appreciate the sophistication of current AI technologies and gain insights into how they're set to evolve.

AI in Your Pocket: How Smartphones Use AI Unknowingly

Facial Recognition and Security

Facial recognition technology has transitioned from a high-tech novelty to a daily security tool for millions. Embedded in your smartphone, AI-driven facial recognition systems are a testament to the advanced capabilities of machine learning. When you set up facial recognition, your phone uses its camera to capture a detailed map of your face, noting unique geometric patterns and landmarks, such as the distance between your eyes or the contour of your cheekbones. AI algorithms then convert this data into a digital model that is as unique as a fingerprint. Each time you unlock your phone, the AI system quickly captures another image, compares it to your registered model, and grants access if the pictures match. This process involves complex AI models known as convolutional neural networks, which are specifically good at picking out patterns in visual data, ensuring that the system is swift and accurate.

Predictive Text and Autocorrect Features

AI also enhances how we communicate on our smartphones. Predictive text and autocorrect features are powered by machine learning algorithms that analyze the words you type and the context in which you type them. By understanding the typical structure of sentences and the relationships between different words, AI can predict the next word you might type or correct a misspelled word to what it thinks you meant. This functionality not only speeds up typing but also makes it more efficient. For instance, if you frequently email your colleague "Jane," your phone's AI will suggest her name once you type "Ja." These systems continuously learn and adapt based on your words, becoming more accurate and helpful the more you type.

Camera Enhancements

AI has dramatically transformed smartphone photography by automating complex decisions that professional photographers spend years mastering. Through AI, your phone's camera analyzes the environment's lighting conditions, the subjects' positions within the frame, and numerous other factors to automatically adjust settings such as exposure, focus, and color balance. Some smartphones even use AI to identify the subject of your photo—be it food, a landscape, or a portrait—and adjust settings accordingly for the best shot. Advanced AI algorithms can also stitch together multiple shots taken at different exposure levels to create a single photograph with enhanced detail in bright and dark areas, a technique known as HDR (High Dynamic Range) imaging.

App Personalization

AI's role extends beyond photography and communication and personalizes your smartphone experience. By analyzing your usage patterns—like the apps you use most often and the times of day you use them—AI algorithms work to personalize app recommendations and advertisements displayed to you. This enhances your user experience by highlighting content and apps that align with your interests and helps app developers and companies improve their services. For instance, if you often use fitness apps in the morning, your phone might suggest health-related content or apps before your workout. All this is done while maintaining a crucial balance with privacy, as modern AI systems are designed to process much of this information directly on your device rather than sending data externally.

Through these examples, it becomes clear just how integrated AI is in the everyday functions of your smartphone, making complex technologies accessible and valuable for your daily life. By transforming raw data into actionable intelligence, AI allows you to interact with your digital world in seamless and personalized ways. As we continue interacting with these intelligent systems, they become more attuned to our preferences and behaviors, promising even greater personalization and efficiency in the future.

Smart Homes and AI: Beyond Voice Commands

The modern home is quietly being transformed by artificial intelligence, far beyond the ubiquitous voice commands from smart speakers. AI in smart homes is becoming increasingly sophisticated, orchestrating a symphony of devices and systems designed to enhance efficiency, security, and accessibility. From managing energy consumption to predicting maintenance needs, AI technologies are adding convenience and improving the sustainability and safety of our living environments.

Energy Management

Imagine your home as an intelligent entity that understands your daily routines, anticipates your energy needs, and adjusts to meet them most efficiently. This is not a glimpse into the distant future but a reality made possible today by AI-driven systems integrated into smart thermostats and lighting systems. These devices learn from your habits—when you wake up, your preferred temperature during the winter, and the rooms you use most—and use this data to optimize heating, cooling, and lighting, reducing unnecessary energy consumption. For instance, a smart thermostat can lower the heating when it detects that the house is empty or the outside temperature has increased. Over time, these adjustments, informed by AI algorithms that analyze and learn from a wealth of data, can significantly reduce energy usage and utility bills. This intelligent energy management extends to larger scales, such as smart grids that use AI to balance energy supply and demand across neighborhoods, further enhancing power use efficiency.

Security Systems

Security is a paramount concern for any homeowner, and AI has revolutionized this aspect of home management by integrating sophisticated surveillance with real-time data processing. Smart home security systems powered by AI enhance safety by distinguishing between ordinary household activity and potential threats. For example, an AI-enabled camera can differentiate between a mail carrier approaching the house and an unknown person lurking suspiciously and only alerting the homeowner or notifying the authorities in the latter scenario. These systems can also integrate with other smart devices in the home to automate responses to perceived threats, such as locking doors or turning on exterior lights, adding a layer of security. Furthermore, the integration with emergency services means that, in the event of an incident, help can be summoned immediately and automatically, often with AI systems providing first responders with real-time data about the situation before they even arrive on the scene.

Maintenance Prediction

AI's predictive capabilities are most strikingly demonstrated in its ability to anticipate maintenance issues within smart homes. By continuously monitoring the performance and output data from smart appliances and systems—like HVAC units and refrigerators—AI can identify patterns or anomalies that may indicate a potential malfunction. This proactive approach to maintenance not only saves homeowners the inconvenience and cost of unexpected breakdowns but also extends the lifespan of the appliances through timely interventions. For instance, if a refrigerator's cooling system begins to show signs of inefficiency, the AI system can alert the homeowner and suggest immediate maintenance, which might involve something as simple as replacing a worn seal or as critical as servicing the compressor, long before the issue leads to a complete breakdown.

Enhanced Accessibility

Beyond convenience and efficiency, AI-driven smart home technology is crucial in enhancing the quality of life for individuals with disabilities. Voice-controlled environments, enabled by AI, allow for seamless interaction with various home systems without physical controls. This can be life-changing for a person with mobility challenges or visual impairments. AI systems can be programmed to respond to voice commands, adjust lighting and temperature, or even operate electronic devices according to the user's specific needs. Moreover, these systems can learn and adapt to the user's preferences, further personalizing the interaction and accommodating the home environment.

As AI continues to evolve, its integration into home environments promises to refine and redefine the comforts of living. By automating routine tasks, providing security solutions, predicting maintenance needs, and enhancing accessibility, AI is setting the stage for homes that are not only smart but also intuitive to the needs of their inhabitants. This integration of AI in home management is a testament to the technology's potential to improve individual lifestyles and the collective experience of community and safety in residential spaces. As we continue to innovate and integrate these technologies, the concept of a "home" will undoubtedly continue to evolve, becoming ever more aligned with the ideals of sustainability, security, and inclusiveness.

AI in Healthcare: Personal Health Monitoring

The integration of artificial intelligence in healthcare is a testament to how modern technology can enhance our understanding and management of health, from routine monitoring to critical care. AI's capability to analyze vast amounts of data and learn from patterns is now paving the way for innovations that not only foresee health issues but also tailor medical interventions to individual needs. This section explores how wearable technology, personalized medicine, mental health apps, and AI-driven emergency responses are revolutionizing our approach to health and wellness, making care more proactive, personalized, and accessible.

Wearable Technology

Consider the wristband that tracks your steps; it's likely also monitoring your heart rate and could analyze your sleep patterns. This wearable technology at work is equipped with AI that continuously gathers and processes physiological data. These devices use algorithms to detect deviations from your standard patterns that might indicate health issues, potentially before you even notice symptoms. For instance, an unusual heart rate or a subtle change in your sleep quality could alert you to underlying conditions that require medical attention. This data helps individuals manage their health more proactively and provides valuable insights to healthcare providers. By tracking these metrics over time, AI can identify trends and predict future health challenges, allowing for interventions that could prevent severe conditions. It's a profound shift from reactive to preventative healthcare, focusing on maintaining well-being rather than merely treating illness.

Personalized Medicine

Personalized medicine is another area where AI is making significant strides, moving away from a one-size-fits-all approach to treatment. This practice involves analyzing a patient's genetic profile, which helps predict how they will respond to certain medications or treatments, optimizing their effectiveness and minimizing side effects. AI algorithms can process genetic data and an individual's health history to recommend the most effective treatments. For example, in cancer treatment, AI can help identify which chemotherapy drugs are most likely to be effective for a particular patient based on the genetic markers of their tumor. This not only spares the patient from the trial and error of finding the proper medication but also significantly increases the chances of successful treatment outcomes. As genomic sequencing becomes more accessible and affordable, AI's role in facilitating personalized medicine is expected to grow, offering hope for more targeted and effective treatments across many conditions.

Mental Health Apps

Mental health dramatically benefits from AI, which is often more complex to diagnose and treat than physical health. Mental health apps equipped with AI technology provide personalized support to users by analyzing their inputs and behaviors. These apps can detect patterns indicative of conditions such as depression or anxiety based on user-reported mood logs, physical activity levels, and even speech patterns. By applying natural language processing, AI can interpret subtle cues in a user's text inputs or voice that may suggest psychological distress. These insights enable the apps to deliver personalized therapeutic recommendations, including mindfulness exercises, behavioral therapy techniques, or prompts to seek professional help. The customized nature of these recommendations makes the guidance more relevant and effective, providing users with timely support that might otherwise be unavailable due to barriers such as stigma or access to mental health professionals.

Emergency Response

AI's impact extends into emergency medical services, transforming how urgent health responses are managed. Consider a wearable device that monitors health metrics and detects life-threatening emergencies such as falls or heart attacks. Upon detecting such an event, AI algorithms can instantly alert emergency services, providing them with critical information about the patient's location and medical history. This rapid response can save lives, especially when every second counts. Additionally,

AI systems in emergency call centers can analyze incoming calls to prioritize cases based on urgency, ensuring that the most critical patients receive immediate attention. This capability enhances the efficiency of emergency services and improves survival rates and outcomes for patients experiencing severe health episodes.

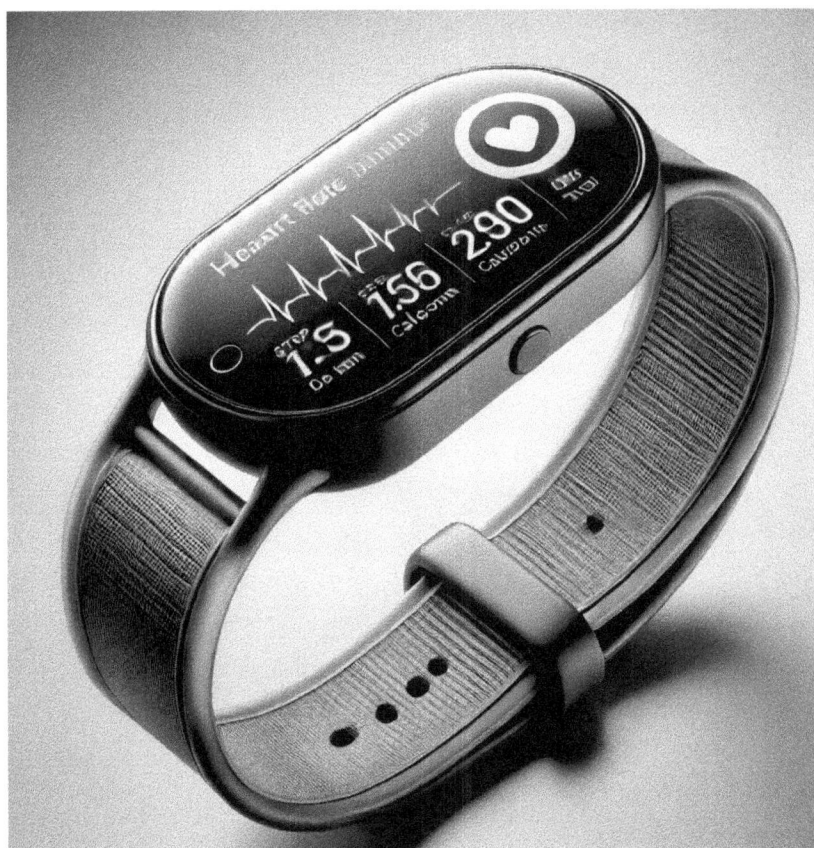

As AI continues to evolve, its integration into healthcare promises profound benefits, making health management more proactive, personalized, and precise. By leveraging the power of AI, we can transform the healthcare landscape, making it more responsive to the needs and nuances of individual patients. This marks a significant advancement in our quest to extend life and enhance the quality of our health and well-being.

Understanding AI in Social Media Algorithms

Social media platforms are not just digital spaces where people connect and share content; they are intricate ecosystems driven by powerful AI algorithms designed to enhance user engagement and streamline content delivery. These algorithms play a crucial role in shaping what you see every time you scroll through your feed, making social media a mirror and molding user preferences and behaviors.

Content Personalization

The core of social media's allure lies in its ability to present you with a feed uniquely tailored to your interests. AI achieves this through sophisticated algorithms that analyze your interactions— what you like, share, comment on, and how much time you spend on different posts. Each action you take is a data point that feeds into the AI, helping it learn your preferences nuancedly. For instance, if you frequently watch cooking videos, the AI notes this pattern and gradually populates your feed with similar content, perhaps from channels you've never seen before. This keeps you engaged and enhances your online experience by aligning the content with your interests. Over time, this capability of AI to adapt and learn from user behavior ensures that the platform evolves continually, becoming ever more effective at keeping users engaged and active.

Ad Targeting

One of AI's most financially crucial applications in social media is in the realm of advertisement targeting. Every day, millions of advertisements are displayed across social media platforms, each aimed at finding the right audience. AI algorithms sort through vast amounts of data—user demographics, browsing histories, purchasing behaviors, and even device types—to display ads most likely to resonate with each user. For example, an AI system might detect that a user has been searching for eco-friendly products. This information will prioritize ads from sustainable brands when curating the user's ad feed. This level of targeting is beneficial for users, who find more relevance in the ads they see. It is also invaluable for advertisers looking to optimize their marketing budgets and increase conversion rates.

Fake News Detection

In an era where information spreads faster than ever, the accuracy and credibility of content can sometimes be compromised. AI is a critical tool in combating the spread of misinformation and fake news. Leveraging natural language processing—a form of AI that helps the system understand and manipulate human language—social media platforms can analyze news articles, posts, and even user comments for signs of fake news. The AI looks for discrepancies in facts, biased language, and questionable sources, among other indicators that might suggest the information is not credible. Once detected, these posts can be flagged, reviewed, and, if necessary, removed, thus maintaining the integrity of the information on the platform. This AI functionality is pivotal in ensuring users have access to reliable information and safeguarding the broader digital information ecosystem.

User Interaction Analysis

Finally, AI's role extends to analyzing user interactions to refine how content is delivered on social media platforms. By examining how users interact with different types of content, AI can identify patterns and preferences that go beyond mere likes or shares. For instance, it might be noticed that certain users prefer short, snappy videos over longer content or that interactive polls keep users engaged longer than static posts. This analysis helps social media companies constantly tweak their algorithms to enhance user experience, ensuring that the content captures attention and encourages more significant interaction. Whether it's by adjusting the visibility of specific posts or by modifying the layout of the feed, AI's continuous analysis of user interactions serves as the backbone for dynamic content delivery that resonates with the ever-changing preferences of the user base.

Through these mechanisms, AI enhances the individual experience on social media and shapes the broader dynamics of how information is consumed and shared globally. By personalizing content, targeting ads, detecting misinformation, and analyzing user interactions, AI is the unseen hand that molds the social media landscape, making it a powerful tool in the digital age. As these technologies evolve, their impact on social media will undoubtedly grow, bringing users worldwide more sophisticated, personalized, and secure online experiences.

The Role of AI in Personal Finance Management

The influence of artificial intelligence has permeated various sectors, significantly transforming operations and enhancing user experiences. Personal finance management is one of the most impactful domains in which AI has made substantial strides. By automating complex processes and providing personalized insights, AI technologies have revolutionized how individuals and businesses manage their financial health. From automated investing to fraud detection and budget management, AI's role in financial services simplifies and amplifies the efficiency and security of financial operations.

Automated Investing

Robo-advisors are one of the most well-known applications of AI in personal finance. These automated investing platforms use sophisticated algorithms to manage investments based on market data and individual investor profiles. By analyzing vast datasets that include historical market trends, global economic indicators, and personal risk tolerance, AI can make informed decisions about where to allocate funds to maximize returns while minimizing risks. For instance, based on user-defined criteria, such as desired retirement age or financial goals, robo-advisors automatically adjust the asset allocation, ensuring the investment portfolio remains robust in varying market conditions. This automation not only democratizes investing, making it more accessible to novice investors without the need for deep financial knowledge, but it also enhances the ability to achieve personalized financial goals more reliably.

Fraud Detection

In an era of ubiquitous digital transactions, the potential for fraudulent activity has risen exponentially. AI systems are crucial in combatting this risk by monitoring transaction patterns and flagging unusual activities that could indicate fraud. These systems analyze every transaction in real-time to detect anomalies, comparing them against established spending patterns and behaviors. For example, suppose there is a sudden high-value transaction from a geographical location that the user has never accessed. In that case, the AI system can flag it and alert the user or temporarily freeze the account to prevent further unauthorized activity. This proactive approach helps prevent financial loss and reinforces the security of users' financial assets. Moreover, as these AI systems learn from new fraud tactics, they continually adapt, thus avoiding fraudulent schemes.

Budgeting Tools

Managing personal finances, especially budgeting, can often be daunting. AI has significantly simplified this task by providing tools automatically tracking and analyzing spending patterns, offering insights and suggestions for better money management. For instance, AI-driven budgeting apps can categorize spending into different buckets, such as groceries, utilities, and entertainment, providing a clear picture of where money goes each month. These tools often offer personalized recommendations for saving opportunities. For example, if the AI notices a monthly subscription that has not been used for several months, it might suggest canceling it to save money. Furthermore, such apps can predict future spending behavior based on past trends, assisting users in making informed decisions about their expenditures before they occur.

Credit Scoring

Traditionally, credit scoring has relied on limited financial indicators, primarily focusing on loan repayment histories and existing credit usage. However, AI introduces a more nuanced approach by analyzing various factors, including transaction history, savings patterns, and social media behavior. This comprehensive analysis allows for a more accurate assessment of a person's creditworthiness, often benefiting those who traditional methods might have underserved. For example, someone without a significant credit history but with a consistent record of responsible financial behavior as indicated by their transaction and savings patterns might score better under an AI-driven system. This more detailed evaluation helps individuals gain better access to credit and enables financial institutions to manage risks more effectively.

These advancements in AI-driven personal finance management streamline financial operations and empower individuals with insights and tools previously available only to financial professionals. By integrating AI into personal finance, users can enjoy enhanced control over their financial health, leading to better economic outcomes and greater peace of mind. As AI technology evolves, its potential to further transform the financial landscape remains vast, promising even more sophisticated tools and systems to manage personal finance efficiently and securely.

TWO

AI in Education—Transforming Learning and Teaching

I magine a classroom where every student receives a uniquely tailored education that adapts seamlessly to their learning pace, style, and interests. This vision is becoming a reality by integrating AI into educational environments. AI's profound impact on education reshapes traditional teaching methodologies and makes learning more personalized, efficient, and inclusive. This chapter explores how AI-driven technologies create customized learning paths in schools, revolutionizing how educators approach teaching and students experience learning.

AI-Driven Personalized Learning Paths in Schools

Customized Learning Experiences

The cornerstone of AI's impact on education lies in its ability to create highly personalized learning experiences. Unlike the one-size-fits-all teaching model, AI algorithms analyze individual student data to understand and adapt to each learner's unique needs. For instance, AI systems can track students' progress across various subjects, identify strengths and weaknesses, and tailor the content accordingly. A student who excels in mathematics but struggles with reading might receive additional literacy resources while still being challenged with advanced math problems. This level of customization ensures that students are neither bored with content that's too easy nor overwhelmed by content that's too hard, enhancing both understanding and retention of knowledge.

Moreover, AI-enabled platforms can incorporate various learning modalities, such as visual, auditory, and kinesthetic, aligning educational content with each student's preferred learning style. This approach increases engagement and promotes deeper learning by leveraging the most effective strategies for each individual. The adaptive nature of AI-driven learning creates a dynamic educational environment where each student can thrive, supported by technology that understands and responds to their evolving educational needs.

Dynamic Assessment Tools

Assessment is essential to the educational process, providing insights into students' understanding and progress. AI revolutionizes this aspect by introducing dynamic assessment tools that evaluate student performance in real-time. These AI systems can adjust the difficulty of tasks based on the student's performance, ensuring that the assessments are both accessible and encouraging. For example, if a student consistently solves math problems correctly, the AI might present more challenging problems to push the student's limits. Conversely, if a student struggles, the AI can dial back the difficulty and provide supportive resources, ensuring that the student remains engaged without feeling frustrated.

This real-time, adaptive assessment allows educators to monitor progress continuously and intervene when necessary, providing targeted support that addresses specific challenges as they arise. By dynamically adjusting to the student's abilities, AI-driven assessments help maintain an optimal learning curve and prevent students from losing interest or confidence in their studies.

Feedback Systems

One of the most significant advantages of AI in education is its ability to provide immediate, actionable feedback. Traditional educational feedback mechanisms often delay task completion and delivery, hindering learning. AI systems, however, can analyze student responses instantly and provide feedback right at the moment of learning, which is crucial for reinforcing concepts and correcting misunderstandings.

This immediate feedback loop enables students to understand what they did right or wrong and adjust their approach accordingly. For instance, if a student mispronounces a word in an AI-driven language learning app, the system can immediately point out the error, provide the correct pronunciation, and even offer tips to improve. This instant feedback ensures students can learn and improve continuously without waiting for teacher intervention. It also frees teachers to focus on in-depth instruction and less on routine grading or basic error correction.

Long-Term Learning Trajectories

AI's ability to collect and analyze long-term data on student performance offers unprecedented opportunities for shaping educational trajectories. By understanding a student's historical data, AI can predict future learning outcomes and suggest personalized learning paths that align with their strengths and career interests. For educators, this insight is invaluable for planning and resource allocation. It allows them to create a learning environment that addresses immediate educational needs and supports students' long-term academic and career goals.

For example, suppose AI analysis reveals a student's strong analytical skills and an interest in science. In that case, educators can encourage this inclination by recommending advanced courses in physics and mathematics or even suggesting extracurricular activities such as robotics clubs or science fairs. This proactive approach to education, powered by AI, helps students build confidence and competence in fields they are most likely to succeed in and enjoy, enhancing their educational and professional prospects.

Through these advanced AI-driven strategies, the educational landscape is transforming, making learning more personalized than ever before. As AI technologies evolve and become more integrated into educational systems, they promise to unlock new potentials for how education is delivered and experienced, paving the way for a future where every student has the tools and support they need to succeed.

Enhancing Classroom Engagement Through AI Tools

The digital age has ushered in a new era of educational tools, with AI at the forefront of transforming traditional learning environments into dynamic and engaging spaces. By incorporating augmented reality (AR) and virtual reality (VR), AI is not just a supplementary tool but a transformative element that enhances how subjects are taught and understood, bringing abstract concepts to life. For instance, imagine a history class where students can use VR headsets to walk through ancient civilizations, interacting with the environment and experiencing history as if they were part of it. This immersive form of learning, powered by AI, makes education an engaging and experiential activity, significantly enhancing student interest and retention of information.

AR and VR, when integrated with AI, provide a platform for creating highly personalized learning experiences. These technologies allow students to explore complex systems or mechanisms in a virtual 3D space, such as dissecting a virtual human body in a biology class or manipulating geometric shapes in a math lesson. The AI component applies to the learner's pace, providing customized guidance and adjusting the complexity of the scenario based on the student's responses and interactions. This tailored approach ensures that students are not overwhelmed by the complexity of the subject matter but are continuously challenged, thereby maintaining engagement and promoting a deeper understanding of the content.

The gamification of learning is another area where AI is making significant inroads. By integrating game mechanics into educational content, AI transforms learning from passive to active. Students earn points, level up, or receive badges as they progress through lessons, which serves to motivate and engage them. These game-like elements make learning more enjoyable, encouraging students to take initiative and strive for better results. AI enhances this process by analyzing individual learning patterns and providing challenges ideally suited to the student's current level, ensuring that the gaming elements are practical and not just superficial additions.

Furthermore, AI-powered systems are equipped to handle students' real-time queries, ensuring continuous and uninterrupted learning. In a traditional classroom setting, students might have to wait for hours or even days to get responses to their questions, which can hinder learning progress. AI-powered assistants, however, are available around the clock and can provide immediate, accurate answers to a wide range of questions. This instant feedback helps solve doubts and allows students to explore related concepts freely without getting stuck or losing interest.

Moreover, AI systems can perform behavioral analytics to identify struggling or disengaged students. By analyzing data points such as login frequency, assignment submission times, and interaction rates during online sessions, AI can alert teachers to students who may require additional support. This early detection system enables timely intervention, ensuring that all students remain on track and receive the help they need before minor issues become significant obstacles to learning.

These AI-driven tools collectively contribute to a more engaging learning environment that responds to students' needs and behaviors. By leveraging AI's capabilities, educators can create a classroom atmosphere conducive to active learning and participation, where every student has the tools and support necessary to succeed. This dynamic approach to education, facilitated by cutting-edge technology, prepares students for the challenges of the modern world and ignites a lifelong passion for learning.

AI and Special Education: Tailored Support Systems

Individualized Support Plans

In special education, the application of AI brings a transformative approach to creating individualized learning experiences that cater specifically to the needs of students with disabilities. The technology's ability to analyze extensive data on a student's learning habits, challenges, and progress allows for the development of customized educational plans that are both adaptive and inclusive. AI systems can process diagnostic information, academic achievements, and continuous feedback from classroom interactions to craft detailed profiles that guide the learning process. For students with special needs, this might mean modifying lesson plans to include more visual aids for those with difficulty with text-based information or providing alternative resources that align with a student's specific learning obstacles.

For example, consider a student with dyslexia who struggles with traditional reading assignments. AI can be programmed to identify such challenges and adapt the learning material accordingly by suggesting multimedia content or interactive activities that reduce reliance on extensive reading and offer alternative methods for comprehension and engagement. This adaptive learning environment helps address each student's unique educational requirements. It promotes a sense of accomplishment and self-worth by allowing students to learn at their own pace and most effectively.

Speech and Language Therapy Tools

AI's impact extends significantly into developing speech and language therapy tools, revolutionizing how students with speech and language impairments are taught and supported in educational settings. AI-powered applications can assist in diagnosing and treating a range of speech disorders, providing engaging ways to facilitate communication skills. For instance, AI speech recognition technologies can accurately identify speech patterns and offer real-time corrective feedback, helping students improve their pronunciation and fluency. Moreover, these tools often include engaging user interfaces and game-like elements that encourage regular practice without the stigma or pressure that might come from more traditional therapeutic settings.

These technologies directly support students and serve as invaluable resources for therapists and educators. They allow for more frequent and consistent practice than might be feasible in a traditional therapy setting, where resources can be limited, and therapy sessions may be restricted by time and scheduling constraints. Additionally, the data collected by AI tools can be used to track progress over time, providing therapists with detailed insights into a student's improvement and areas that require more focused attention.

Motor Skills Development

AI is also making strides in developing motor skills, mainly through robotics engineered to assist students with physical disabilities. These robotic systems are designed to offer repetitive practice and personalized assistance, which are crucial in developing fine motor skills. Through interactive tasks and adaptive feedback, AI-driven robots can engage students in exercises that improve coordination, strength, and agility. For example, a robotic arm with AI software can adapt responses based on the student's movements, assisting as needed while encouraging independent effort and gradual improvement.

These AI-assisted robotic systems can be particularly transformative in unique education settings where traditional methods might not fully meet students' diverse needs. The technology offers more personalized and intensive support and incorporates elements of play and interaction, making the learning process more enjoyable and less daunting for students struggling with physical challenges.

Emotional Recognition

One of the most innovative applications of AI in special education is the use of emotional recognition technology. By analyzing vocal tones, facial expressions, and body language, AI systems can gauge the emotional state of students and adjust the learning environment accordingly. This capability is significant for students who may have difficulties expressing their feelings or might not respond well to traditional disciplinary approaches. For instance, if a student becomes frustrated during a learning activity, AI can detect this emotion and modify the task to make it less challenging, or it might suggest a break or a different activity that could help calm the student.

Furthermore, emotional recognition AI can help educators under-stand subtle cues that indicate a student's emotional state, enabling more effective and empathetic interactions. This fosters a more supportive learning environment and helps build trust and confi-dence between students and their teachers. As emotional aware-ness is crucial in special education, AI technologies that can interpret and respond to these emotional signals are essential in creating inclusive classrooms that cater to all students' emotional and educational needs.

Through these diverse applications, AI is significantly enhancing the landscape of special education. By providing tailored educa-tional plans, innovative therapy tools, interactive motor skills development, and sensitive, emotional recognition, AI technolo-gies offer new pathways for students with disabilities to achieve their academic goals. This revolutionizes how educational content is delivered and adapted and ensures that all students have the support and resources they need to succeed in their educational journeys.

Evaluating AI Educational Software: A Guide for Teachers

When considering the integration of AI into educational environ-ments, selecting the right software becomes a pivotal decision for educators. The criteria for choosing AI educational tools encom-pass several vital factors: user-friendliness, alignment with curriculum standards, and data security. Each aspect is critical in ensuring that the software enhances the learning experience, aligns seamlessly with educational goals, and protects sensitive information.

User-friendliness is essential because it determines how easily students and teachers interact with the software. A difficult-to-navigate or-understand tool can become a barrier to learning rather than a facilitator. The AI software must have an intuitive interface that you and your students can use without extensive training. Moreover, the software's ability to integrate smoothly with existing systems and software used by the school can significantly impact its effectiveness and uptake. For instance, an AI-based learning platform that easily syncs with the school's learning management system (LMS) allows for a smoother transition and reduces the learning curve for students and faculty.

Alignment with curriculum standards is another critical criterion. The AI tool should adhere to and enhance the curriculum by providing resources and learning experiences directly tied to the educational standards and outcomes expected at each grade level. This ensures that the use of technology is purposeful and directly contributes to the academic objectives. For example, AI software that offers math tutoring should align with the common core standards, providing practice and instruction that reinforce what is taught in the classroom.

Data security is paramount when implementing any technology in schools. AI educational tools often require access to sensitive student data, including learning patterns, personal information, and academic records. The software must comply with all relevant privacy laws, such as FERPA in the United States, which protects student education records. The AI provider must demonstrate robust security measures to protect data from unauthorized access and breaches. Transparency about how data is collected, used, and stored should be communicated to the school administrators and the parents, ensuring everyone is informed, and any concerns are addressed proactively.

The effectiveness of AI educational tools can be measured through specific performance metrics that reflect improvements in student engagement and academic performance. Educators can gain insights into the software's impact on learning by analyzing how students interact with the AI tool. Metrics such as time spent on tasks, completion rates, and progress in specific skills are invaluable for evaluating the effectiveness of the technology. Additionally, changes in student engagement levels can be monitored through data on logins and active use, providing a quantitative measure of how the AI tool influences students' willingness to participate in the learning process.

Ethical considerations also play a significant role in using AI in education. One of the primary concerns is the potential for bias in AI algorithms, which can manifest in the tools favoring specific student demographics over others. It is crucial that AI software is designed and continually evaluated to ensure it promotes equity and inclusivity. Educators should be vigilant about any bias in how the software interacts with students or evaluates their work and be prepared to intervene if necessary. Additionally, the ethical use of AI ensures that the technology does not replace the human element in education but enhances the teacher-student interaction, complementing the educators' roles and supporting tailored learning experiences.

Continuous improvement is essential to ensure that AI tools remain practical and relevant. AI technology evolves rapidly, and educational software must be regularly updated to incorporate the latest advancements and improvements. This involves technical updates and adapting the learning content and tools based on feedback from educators and students. Regular assessments of the AI software can provide insights into areas where it excels and where improvements are needed, guiding further development. Additionally, ongoing training for teachers on fully embracing AI tools to maximize the benefits for educators and students ensures the technology is used to its full potential.

By carefully selecting AI educational software that meets these criteria, educators can enhance their teaching toolkit with powerful resources that enrich students' learning experiences. Integrating AI into education offers exciting possibilities, but thoughtful consideration is required to ensure it is a beneficial addition to the educational landscape. As AI continues to develop, its role in education will undoubtedly expand, providing new ways to engage students and personalize learning to meet the needs of every learner.

Preparing Students for an AI Future: Curriculum Strategies

In a world increasingly driven by artificial intelligence, equipping students with AI literacy has become as fundamental as reading, writing, and arithmetic. Integrating AI literacy into the curriculum isn't just about understanding algorithms or coding; it's about ensuring students grasp the profound implications of AI across various sectors and how they can harness this knowledge for future innovations. To embed AI literacy effectively, educators can start by demystifying AI technologies, explaining in simple terms how they work and are applied in everyday life—from the algorithms that filter their social media feeds to the AI mechanisms that predict their online shopping preferences. This foundational knowledge can then be expanded to include discussions about more advanced AI applications in healthcare, automotive, and finance, illustrating the breadth of AI's impact.

Further, it's crucial to incorporate structured lessons exploring AI systems' design and ethical considerations. For instance, students could evaluate case studies that highlight both successful AI implementations and those that have failed due to ethical oversights. This dual approach ensures students understand AI's capabilities, limitations, and the importance of ethical considerations in its application. To further this learning, schools can leverage existing resources like online platforms offering AI education modules or partner with tech companies providing expert guest lectures and interactive AI demonstrations.

As the job landscape evolves, the demand for skills like problem-solving, critical thinking, and digital literacy has surged. These competencies are essential for students to thrive in an AI-driven world. Problem-solving skills can be enhanced by engaging students in scenarios that require them to utilize AI tools to find solutions to real-world problems, such as climate change or logistical challenges. Critical thinking can be fostered by encouraging debates on the implications of AI in decision-making processes, pushing students to consider various perspectives and the potential biases of AI systems. Digital literacy should go beyond basic computing skills to include understanding data privacy, cybersecurity, and the ethical use of digital tools. Schools can support the development of these skills by creating maker spaces or tech labs where students can experiment with AI software and hardware under guided supervision to learn by doing.

Collaborative projects that involve AI tools can significantly enhance hands-on learning and innovation. Educators can facilitate projects where students use AI to create solutions, such as developing a simple chatbot or using machine learning models to analyze data collected in science experiments. These projects allow students to apply their AI knowledge in practical settings and promote teamwork and creativity as students collaborate to solve problems and share their findings. The key is providing students access to AI tools in a controlled environment where they can experiment safely and learn from their successes and failures.

Teaching the ethical use of AI forms a critical pillar of AI literacy. Students must understand that AI can drive immense technological advancements and pose significant ethical and societal risks. Educators should introduce concepts like algorithmic bias, surveillance, and the socio-economic impacts of automation. Discussions can be structured around current events or historical examples of the technological effects on society to make these issues relatable and tangible. Furthermore, educators can encourage students to consider their role in developing or using AI ethically, fostering a sense of responsibility and agency. This aspect of AI education is not just about caution but about empowering students to contribute to a future where technology enhances societal well-being.

By addressing these critical areas in AI education, educators can provide students with a robust toolkit of knowledge and skills essential for navigating the future. These efforts prepare students for the technical demands of future careers and instill an ethical framework that will guide their decisions as future innovators.

As this chapter concludes, we reflect on AI's transformative potential in education. From personalizing learning to equipping students with the skills to excel in a tech-driven future, AI is not just a subject to be taught but a significant educational tool. As we progress into the next chapter, we'll explore how AI reshapes other sectors, further underscoring the importance of integrating AI literacy into academic curricula. By preparing students today, we pave the way for a more informed, ethical, and innovative tomorrow.

THREE

Ethical AI–Balancing Innovation With Responsibility

In the tapestry of modern technological advancements, artificial intelligence (AI) stands out as a paradigm of innovation, weaving complex patterns of efficiency and opportunity into the fabric of daily life. Yet, as we integrate these powerful capabilities more deeply into our personal and professional spheres, we must also navigate the intricate ethical landscape accompanying them. This chapter explores the delicate balance between leveraging AI to enhance our lives and ensuring that we do so responsibly, with a firm commitment to upholding ethical standards that protect privacy, ensure fairness, and maintain our human values.

Privacy Concerns and AI: Finding the Balance

Data Collection Practices

In the digital age, personal data acts as currency in a complex economy of information exchange and powering systems that make life more convenient, connected, and tailored to your needs.

However, the mechanisms of AI-driven data collection often operate beneath a veil of opacity, leaving many unaware of how deeply these algorithms reach into their private lives. Transparency in AI-driven data collection is a regulatory requirement and a cornerstone of trust between technology providers and users. It involves clear communication about what data is being collected, how it is used, and why it is necessary. For instance, when a fitness app uses AI to track your physical activities and suggest health improvements, it should explicitly inform you that it processes data related to your location, physical stats, and workout patterns. This level of transparency ensures that you, as a user, understand the implications of your data usage and feel secure in how your personal information is handled.

Consent and Control

The principle of consent is fundamental to ethical data practices. AI systems must ask for your permission before gathering your data and empower you to control what you share and how it is used. This means providing clear options for opting out of data collection or deleting your data from the system. Furthermore, these controls should be easily accessible and straightforward, ensuring you can exercise your rights without burden. For instance, a smart home device that uses AI to learn your living patterns should offer a simple interface for you to adjust what information is recorded and to opt out of data collection for certain types of activities. This level of control helps maintain a balance where AI can be used to enhance convenience without compromising your autonomy over your personal information.

Regulatory Compliance

The landscape of AI development is not just shaped by innovators and markets but also by regulators, who play a critical role in ensuring that technology serves the public good. Regulations like the General Data Protection Regulation (GDPR) in the European Union set international benchmarks for data protection and privacy, imposing strict guidelines on how AI can interact with user data. These regulations require AI systems to incorporate privacy by design, meaning that data protection is an integral part of the technology development process, not just an afterthought. By complying with such regulations, developers enhance the security and integrity of AI systems and build trust with users who are rightfully concerned about their privacy. The implications of these regulations extend beyond borders, influencing global AI development standards and encouraging a worldwide commitment to ethical practices in technology.

Privacy-Preserving Techniques

As AI continues to evolve, so do the techniques designed to safeguard privacy. Differential privacy, for instance, is a technique that allows companies to glean valuable insights from datasets while mathematically guaranteeing the anonymity of individual entries. This means that AI can analyze patterns in user data to improve services or products without exposing or risking the privacy of individual users. Another innovative approach is federated learning, a model that enables AI to learn from data stored on your device rather than being uploaded to a central server. This method minimizes the risk of data breaches and reduces the amount of personal data transmitted across the internet. These technologies represent a proactive approach to privacy that aligns with ethical

standards and respects user rights, thus fostering a healthier relationship between technology developers and the public.

Through these discussions, it becomes clear that the ethical
deployment of AI is not merely a technical challenge but a moral
imperative. As we continue to explore artificial intelligence's vast
potential, let us also commit to the principles of responsibility,
transparency, and respect for individual privacy. Doing so ensures
that AI remains a force for good, enhancing our capabilities
without compromising our values.

Bias in AI: Causes, Consequences, and Mitigations

The allure of artificial intelligence lies in its ability to analyze vast
amounts of data and make decisions at speeds and accuracies that
humans cannot match. However, this strength is also its Achilles'
heel when the data it learns from or the algorithms that power it
are biased. Bias in AI can originate from various sources, each
potentially skewing the technology's outputs in ways that can have
real-world consequences. The data used to train these systems is
one primary source of AI bias. The AI will likely perpetuate these
biases if the data reflects historical inequalities or lacks diversity.
For example, suppose an AI system is trained on job application
data from a tech industry that has historically been male-domi-
nated. In that case, it may inadvertently favor male candidates, not
because they are more qualified, but because there are more
historical data points for males. Another source of bias is the
design of the algorithm itself. Algorithmic design choices, such as
which variables to include and how to weigh them, can introduce
bias, often subtly and unintentionally. If a loan approval AI gives
higher weight to credit scores and the credit scoring system itself
is biased against specific demographics, the AI will replicate and
possibly amplify this bias.

The impacts of biased AI systems are profound, especially when deployed in critical areas such as hiring, law enforcement, and financial services. In hiring, a biased AI system might filter out qualified candidates from underrepresented groups, denying opportunities based on gender, race, or age. In law enforcement, AI tools used for predictive policing could disproportionately target specific communities, leading to over-policing and an erosion of trust in these areas. In loan approvals, biased AI could deny financial opportunities to individuals based on flawed criteria linked to their background rather than their actual credit-worthiness. These biases do not just perpetuate existing inequalities; they can deepen them, embedding unfair practices into the systems that many hope would be forces for good.

Detecting and correcting bias in AI systems is both crucial and challenging. One effective methodology is using audits and fairness metrics to assess AI outputs. Organizations can identify potential biases in their models by applying statistical tests to evaluate whether AI decisions are equitable across different groups. For instance, if an AI hiring tool shows a significant discrepancy in the rate of recommendations between male and female candidates despite similar qualifications, it likely indicates a bias that needs to be addressed. Correcting this bias might involve revisiting the training data to ensure it is representative and unbiased or adjusting the algorithm to mitigate the influence of biased factors. Another approach is to use synthetic data to balance the datasets or to simulate data for underrepresented groups to train the AI more equitably.

Continuous monitoring and evaluation of AI systems are essential to ensure biases are addressed and kept in check over time. AI systems are not static; they learn and evolve as they are fed new data. Without ongoing oversight, a system that starts bias-free can develop biases as it operates. Regularly revisiting the AI models with updated audits and refining the algorithms as necessary helps maintain their fairness and efficacy. Setting up dedicated teams within organizations to oversee AI fairness and ethics can institutionalize these efforts, ensuring that they are not an afterthought but an integral part of AI development and deployment processes.

By actively engaging in these practices, we can harness AI's benefits while minimizing its risks, ensuring that it acts as a tool for promoting equity rather than an instrument for entrenching bias. This proactive approach to managing AI bias enhances the technology's effectiveness and builds public trust in how these advanced systems are employed across various sectors. As AI permeates every aspect of our lives, our commitment to addressing and mitigating its biases will be paramount in shaping a future where technology equitably benefits all.

AI in Law Enforcement: Ethical Boundaries

Surveillance and Privacy

In law enforcement, deploying AI technologies, particularly in surveillance, heralds a new era of public safety and crime prevention. However, this advancement brings significant ethical concerns, primarily regarding the balance between enhancing security and protecting individual privacy rights. AI-enabled surveillance systems, such as facial recognition cameras and predictive policing tools, are incredibly efficient in gathering and analyzing data to identify potential threats and prevent crimes. While the benefits of such capabilities in terms of heightened security are undeniable, they also raise critical questions about how surveillance should penetrate the public's private lives.

Consider the scenario where AI surveillance tools are used in public spaces to track individuals deemed suspicious by law enforcement agencies. While this might aid in preventing potential criminal activities, it also risks the privacy of countless innocents whose data is collated and analyzed without their consent. The ethical dilemma is determining where to draw the line between public safety and personal privacy. Should a person's every move in public spaces be open to scrutiny simply because technology has made it possible? The key is to develop regulatory frameworks that govern the use of AI in surveillance, ensuring that while public spaces are monitored for safety, the data collected is used to respect and preserve individual privacy. This involves setting limits on what data can be collected and how long it can be stored and implementing robust oversight mechanisms to monitor how law enforcement uses AI tools, ensuring they stay within bounds.

Decision Support Systems

AI's role as a Decision Support Tool in law enforcement can significantly enhance the efficiency and accuracy of police work. By analyzing data from various sources, AI can help make informed decisions regarding resource allocation, response strategies, and even predictive assessments of crime hotspots. However, relying on AI for critical decision-making in law enforcement also introduces risks, mainly when decisions are made without sufficient human oversight. The danger lies in AI systems making erroneous recommendations based on flawed data or algorithms, which can lead to unjust outcomes.

For example, the stakes are incredibly high if an AI system is used to assess the risk level of individuals on probation and recommend whether they should be detained or released. An incorrect assessment due to biased data can lead to someone being wrongfully imprisoned or, conversely, a high-risk individual being released. It is, therefore, crucial to maintain a system where AI recommendations are constantly reviewed by human officers who can consider broader contexts and ethical implications. This hybrid approach ensures that the efficiency of AI is harnessed to improve law enforcement practices while safeguarding against the risks of automated decision-making processes.

Accountability in AI Deployments

The deployment of AI in law enforcement must be accompanied by stringent accountability measures to ensure the technology is used responsibly. Every decision AI makes, especially those affecting individuals' lives and freedoms, must be traceable and justifiable. This is crucial not only for the sake of transparency but also for maintaining public trust in law enforcement agencies. Accountability in AI deployments can be achieved by keeping comprehensive logs of AI decisions and the data used to reach those decisions, which can be audited if discrepancies or disputes arise.

Moreover, law enforcement agencies should be open about their use of AI technologies, including the capabilities and limitations of these systems. By keeping the public informed, agencies can prevent misunderstandings and resistance and promote a cooperative relationship with the community. Clear policies should also be established to outline the acceptable uses of AI in law enforcement, and regular reviews should be conducted to ensure these policies are adhered to. This proactive approach to establishing and maintaining accountability prevents misuse of AI and ensures that its deployment in law enforcement contributes to the fair and just administration of the law.

Community Engagement

Gaining the trust and acceptance of the community is essential for successfully deploying AI technologies in law enforcement. When communities are involved in discussions about how AI is used in policing, it fosters a sense of cooperation and mutual respect. Engagement initiatives can include public forums, workshops, and consultations where community members can voice their concerns and suggestions regarding the use of AI in law enforcement. Such interactions not only educate the public on the benefits and risks of AI but also provide law enforcement agencies with valuable insights into community sentiments, which can guide the development and implementation of AI technologies.

For instance, if a police department plans to implement an AI-driven surveillance system in a neighborhood, it should first seek input from residents. Understanding their perspectives on safety, privacy, and technology can help tailor the system to meet the community's needs while respecting its concerns. This collaborative approach enhances the effectiveness of AI deployments and ensures they are conducted in a manner that respects and upholds the community's values and rights. By prioritizing transparency and collaboration, law enforcement agencies can leverage AI to enhance public safety and build more robust, trusting relationships with the communities they serve.

The Role of AI in Sustainable Development

In the quest for sustainability, AI emerges as a pivotal ally, transforming how we manage and conserve our planet's precious resources. The integration of AI in resource management, notably in the water and energy sectors, showcases a path toward preserving these vital assets and optimizing their use to support a sustainable future. For instance, AI-driven systems in water management can predict usage patterns and detect leaks in real-time, significantly reducing wastage and ensuring efficient distribution. These systems analyze vast datasets gathered from sensors across the water network, identifying anomalies that could indicate leaks or inefficiencies, and automatically adjust control mechanisms to optimize water flow and pressure. Similarly, AI algorithms can forecast power demand and supply fluctuations in energy management, enabling more competent grid management. By integrating data from weather forecasts, consumer behavior patterns, and energy production statistics, AI helps utility providers dynamically adjust power generation and distribution, reducing waste and enhancing the use of renewable energy sources.

This intelligent management extends into renewable energy sectors, such as wind and solar power, where AI's predictive capabilities are used to anticipate energy production levels based on weather conditions. This allows for better integration of these somewhat unpredictable energy sources into the power grid, ensuring that the maximum amount of renewable energy can be harnessed without compromising the stability of the energy supply. The result is a more resilient, efficient, clean energy system, moving us closer to our sustainability goals. Moreover, AI's role in optimizing resource use also has significant economic implications, as enhanced efficiency translates into cost savings and prolonged asset life, which is crucial for the financial sustainability of utilities and communities.

The looming threat of climate change presents the most daunting challenge to global sustainability efforts. Here, too, AI stands as a crucial tool in our arsenal. Through its advanced data processing capabilities, AI aids in monitoring climate change impacts with unprecedented precision and scale. Satellite images, atmospheric data, and oceanographic sensors feed into AI systems that track weather patterns, ice cap volumes, and sea level changes. This detailed monitoring enables scientists to make more accurate predictions about the pace and impact of climate change, informing policy decisions and public awareness initiatives.

Furthermore, AI plays a crucial role in developing strategies for mitigating and adapting to climate change. For mitigation, AI algorithms optimize energy systems to reduce emissions, enhance the efficiency of transportation networks, and improve waste management practices. Regarding adaptation, AI helps design more innovative infrastructure capable of withstanding extreme weather events and rising sea levels, ensuring that communities are better prepared to face the already underway changes.

Transitioning to sustainable agricultural practices is essential to meeting the global food demands of a growing population while minimizing environmental impacts. AI technologies usher in a new era of precision agriculture, where every aspect of farming, from sowing seeds to harvesting crops, can be optimized for sustainability. AI-driven systems analyze data from field sensors, drones, and satellites to monitor crop health, soil conditions, and weather patterns, providing farmers with precise information on where to water, fertilize, or apply pest control. This targeted approach boosts crop yields and drastically reduces water usage, chemical runoff, and greenhouse gas emissions from agricultural machinery. The ability of AI to process and analyze the vast amounts of data generated on modern farms makes it possible to implement these practices at scale, transforming large agricultural operations into sustainability models.

In the industrial sector, the drive for economic sustainability is well complemented by AI's capabilities in optimizing manufacturing processes. AI systems streamline production lines, reduce waste, and improve product quality by continuously analyzing data from the manufacturing process and making real-time adjustments. This not only lowers the cost of manufacturing but also extends the lifespan of machinery and reduces the industry's environmental footprint. AI's impact extends beyond the factory floor, with sophisticated algorithms optimizing supply chain logistics, reducing unnecessary transportation and associated emissions, and ensuring products are delivered more efficiently to markets. This holistic approach to industrial sustainability showcases AI's potential to foster environments where economic growth does not come at the expense of ecological health.

As AI continues to evolve and expand its capabilities, its role in promoting sustainable development across multiple domains is transformative and vital. By leveraging AI, we can ensure that our pursuit of economic and social progress does not undermine the ecological balance upon which all life on Earth depends. In this way, AI is a tool for innovation and efficiency and a cornerstone of our collective efforts to secure a sustainable future for future generations.

AI and Accessibility: Creating Inclusive Technologies

The imperative for inclusivity in technology design extends profoundly into artificial intelligence. As AI systems become more integral to everyday life, ensuring these technologies are accessible to people with disabilities is not just a consideration—it's a necessity. Designing for accessibility means creating AI solutions that can be used as effectively by people with disabilities as by those without. This approach should permeate every phase of AI development, from the initial design concept to user interface choices and beyond. For example, voice-activated AI assistants that can understand and process spoken commands in natural language are invaluable for individuals who cannot use traditional computing devices due to physical limitations. Similarly, AI-driven applications that adjust their output based on user interaction patterns can help accommodate users with cognitive impairments by simplifying navigation or enhancing the visibility of critical features according to individual needs.

The role of AI in enhancing assistive technologies has seen significant advancements, particularly in the development of visual and auditory aids. For visually impaired users, AI-powered applications can transform text into speech or provide descriptive audio for images and videos, thus bridging the gap between visual content and comprehension. In auditory assistance, AI-enhanced hearing aids can now filter and amplify sounds, making it easier for users with hearing impairments to engage in conversations in noisy environments. These technologies not only improve the quality of life for individuals with disabilities but also foster greater independence by enabling more seamless interaction with the digital world.

Despite these advances, several barriers still prevent the widespread accessibility of AI technologies. One major challenge is the lack of awareness and understanding of accessibility needs within the tech community, which can lead to the oversight of these requirements in the design process. Additionally, economic barriers often prevent the full integration of accessibility features, as prioritizing these can increase development costs. Overcoming these challenges requires a concerted effort from developers, companies, and policymakers to prioritize inclusivity and invest in developing universally accessible AI technologies.

Ethical considerations are paramount when discussing AI and accessibility. The development of AI systems that exclude or fail to adequately serve individuals with disabilities not only perpetuates existing inequalities but also runs counter to the ethical principle of fairness. Ensuring that AI systems promote inclusivity involves more than just technical adjustments; it requires a fundamental commitment to universal design—a philosophy that advocates for designing products and environments usable by all people to the greatest extent possible without needing adaptation. By adhering to this philosophy, AI developers can create technologies that do not discriminate, consciously or unconsciously, against users with disabilities, thereby upholding the values of equity and access.

Integrating AI into accessibility efforts represents a significant step forward in building a more inclusive society. By embracing the principles of universal design and continuously striving to overcome barriers to accessibility, we can ensure that AI technologies reach their fullest potential, not just in terms of innovation but also in enabling all individuals to lead more prosperous, more autonomous lives. This commitment to inclusivity enriches the field of AI and reflects our broader societal values of diversity and equality, reinforcing the role of technology as a force for good. As we continue to explore the vast possibilities of artificial intelligence, let us remain steadfast in our dedication to creating technologies that uplift every member of our global community, regardless of their physical or cognitive abilities.

In conclusion, this chapter underscores the critical importance of embedding inclusivity into the fabric of AI development. From enhancing personal independence with tailored assistive technologies to advocating for ethical practices that encompass all users' needs, AI promises a more accessible world. As we turn the page to the next chapter, we will delve into how AI is shaping the future of healthcare, exploring its potential to transform medical care and the ethical implications accompanying these advancements. By prioritizing inclusivity and responsibility, we can harness the full potential of AI to serve the common good and foster a society that values and supports every individual.

FOUR

AI at Work—Enhancing Productivity and Innovation

I n the bustling arena of modern workplaces, where deadlines loom and projects jostle for attention, efficiency isn't just a goal—it's a necessity. Enter artificial intelligence (AI), your unseen but ever-diligent assistant, ready to streamline operations, predict risks, and facilitate clear communication within project teams. This chapter delves into how AI is transforming the landscape of project management, turning potential chaos into structured success. As you navigate this discourse, imagine AI not merely as a tool but as a strategic partner in orchestrating the complexities of project management with a level of precision and foresight previously unattainable.

AI in Project Management: Streamlining Operations

Automated Task Management

One of the cornerstone benefits of AI in project management is its ability to automate routine tasks, liberating human minds to focus on areas requiring creative or strategic input. This automation spans various dimensions of project management, including scheduling, resource allocation, and progress tracking. AI algorithms excel in handling these operations by analyzing project timelines, team member availability, and task durations to optimize schedules dynamically. For instance, if an unexpected delay occurs, AI can immediately recalibrate the schedule and reassign tasks based on current priorities and team workload, ensuring project milestones are met. This boosts efficiency and enhances team productivity by allocating human resources where they are most effective, thus maximizing the return on human capital.

Risk Assessment Models

In any project, the potential for risk looms as a constant shadow. Traditional methods of risk management often involve manual monitoring and gut instincts, which, while valuable, leave room for human error. AI transforms this landscape by deploying sophisticated risk assessment models that analyze historical data and ongoing project metrics to forecast potential pitfalls. These models use machine learning to identify patterns that may indicate an impending risk, from budget overruns to scope creep and provide actionable insights to mitigate these risks proactively. By predicting potential problems before they manifest, AI enables project managers to implement strategic interventions early, safeguarding the project from delays and budget inflations, thus securing both timelines and financial resources.

Enhanced Communication

Clear communication is the linchpin of effective project management. AI enhances this critical aspect by analyzing communications such as emails, instant messages, and meeting notes to ensure that vital information is promptly shared among all stakeholders. Natural language processing, a subset of AI, enables these systems to understand and prioritize information based on relevance and urgency. For example, AI tools can highlight action items from a meeting transcript and automatically generate reminders for team members, ensuring that essential tasks are noticed in the shuffle of daily activities. This capability keeps projects on track and fosters team cohesion by maintaining a clear and consistent flow of information.

Decision Support Systems

At the heart of project management lies decision-making—a complex task with significant implications for project success. AI aids in this process by providing project managers with data-driven insights that inform better decisions. Through the integration of AI, project data is continuously analyzed, offering real-time insights into project health, resource utilization, and performance trends. This information enables managers to make informed decisions about project direction and priorities, such as allocating additional resources to a lagging project phase or adjusting project scope to meet emerging challenges. By equipping managers with a detailed, data-driven view of the project landscape, AI transforms decision-making from a reactive task to a proactive strategy, enhancing the overall efficiency and success of projects.

As AI continues to evolve and integrate more deeply into project management, its role as a catalyst for innovation and efficiency becomes increasingly undeniable. By leveraging AI, organizations can not only enhance the productivity of their teams but also gain a competitive edge in the fast-paced business environment. This chapter sets the stage for a deeper exploration of AI's transformative potential across various industries, heralding a new era of productivity and innovation in the workplace. As we progress, let us continue to explore the myriad ways AI can be harnessed to foster business success and a more dynamic and responsive organizational culture.

AI and Customer Service: Beyond Chatbots

In the dynamic realm of customer service, artificial intelligence (AI) is revolutionizing interactions in ways that extend far beyond the familiar chatbots. Today's AI is responsive and anticipatory, predicting customer needs and enhancing service delivery preemptively. This proactive approach to customer service, powered by AI's predictive analytics, involves a sophisticated analysis of customer behavior patterns. By sifting through vast quantities of data—from past purchases and support interactions to browsing histories—AI can identify and address potential issues before they escalate into real problems. Imagine a scenario where AI detects a customer frequently visiting the help page after purchasing a product. Before the customer even reaches out, AI can trigger a personalized instructional video or an offer for a free consultation session, effectively preventing frustration and enhancing the customer's experience. This boosts satisfaction and fosters loyalty by demonstrating attentiveness and a commitment to user satisfaction that anticipates needs rather than merely reacting to them.

Moreover, AI's capabilities in emotion recognition are transforming how businesses interact with customers on a more personal level. AI tools can now accurately gauge customer emotions by analyzing verbal cues, text sentiment, and facial expressions during video interactions. This emotional intelligence allows customer service systems to tailor responses to the customer's emotional state, ensuring that communications are effective and empathetic. For example, if a customer expresses frustration, the AI system can adjust its tone, perhaps by softening language or by escalating the query to a human agent informed about the customer's emotional state and the interaction context. This nuanced approach to customer service, sensitive to the emotional undercurrents of interactions, significantly enhances the quality of service, turning routine transactions into opportunities for meaningful engagement.

Integration with the Internet of Things (IoT) represents another frontier where AI is making significant inroads in customer service. With IoT devices, AI can perform seamless and anticipatory service interventions in homes and offices. Suppose an IoT-connected refrigerator detects a malfunction, for instance. In that case, AI can automatically initiate a service ticket, inform the customer, and schedule maintenance without the customer needing to diagnose the problem or contact support. This integration streamlines the maintenance process and enhances the user experience by reducing downtime and inconvenience. The interconnectedness of IoT devices coupled with AI's analytical prowess means that service can be more responsive, more attuned to the rhythms of daily life, and less intrusive, all of which elevate the standard and expectation of what good customer service can be.

Lastly, continuous learning in AI systems is pivotal in the evolution of customer service. AI systems are designed to learn from every interaction and continuously refine their algorithms to meet customer needs better. This learning process involves analyzing explicit and inferred feedback from customer behavior to improve response accuracy and service quality over time. Each interaction feeds into the AI's understanding, allowing it to predict customer preferences better and personalize interactions. Over time, this results in a customer service system that understands what customers want and adapts to changing preferences and circumstances, ensuring relevance and precision in every interaction. Through continuous learning, AI becomes an ever-more proficient partner in delivering customer service that is both personal and scalable, reflecting the changing dynamics of consumer expectations and technological possibilities.

As AI continues to evolve, its integration into customer service moves beyond mere automation, transforming into a dynamic system that anticipates needs, understands emotions, and learns from every interaction. This enhances efficiency and redefines the quality of service, providing customers with experiences that are increasingly personalized, empathetic, and seamlessly integrated into their interconnected lifestyles. As we explore AI's applications further, it becomes clear that its potential to transform industries is not just about technological advancement but also about crafting more profound, meaningful interactions that resonate on a personal level.

AI in Marketing: Personalization at Scale

In today's digital age, the marketing landscape is evolving more rapidly than ever, driven by an explosion of data and increasingly sophisticated consumer expectations. At the forefront of this revolution is artificial intelligence (AI), which has become a linchpin in understanding and predicting consumer behavior at an unprecedented scale. By diving deep into vast datasets that include everything from social media interactions and browsing histories to transaction records and location data, AI provides marketers with insights that were once beyond reach. This deep understanding enables crafting personalized marketing strategies that resonate individually rather than relying on broad, demographic-based campaigns. For example, AI can identify patterns indicating consumers' preference for eco-friendly products based on their browsing habits and past purchases. Marketers can then tailor their messages to highlight the sustainability features of products, significantly increasing the relevance and effectiveness of their campaigns.

The process of dynamic content delivery further exemplifies how AI is transforming marketing efforts. As you scroll through your favorite online platforms, AI is working subtly in the background, analyzing how you interact with different types of content. This continuous analysis helps AI systems learn your preferences, not just in terms of content topics but also content formats, timing, and presentation styles. Based on this understanding, AI dynamically adjusts what you see, ensuring that the content aligns with your interests and engages you most effectively. For instance, AI will schedule video posts accordingly if you engage more with video content in the evenings. At the same time, your morning feed might feature quicker reads, such as infographics or short articles. This level of personalization ensures that each user's interaction is maximized for engagement, transforming passive viewers into active participants in the digital marketing ecosystem.

Moreover, AI-driven campaign optimization allows marketers to refine their strategies in real-time, adapting to the fluid dynamics of consumer interactions and market trends. Through AI, marketing campaigns are no longer static entities launched into the consumer ether with hopes of performance. Instead, they are dynamic and adaptive, continuously learning from ongoing consumer responses. AI analyzes performance data across multiple channels, identifying what is working and what isn't, and automatically adjusts various campaign elements—from allocating budgets between different platforms to tweaking message tones and visuals. This real-time optimization improves the effectiveness of marketing campaigns and enhances return on investment by allocating resources to strategies that yield the best results.

The advent of voice and visual search technologies has opened new frontiers for marketing, and AI is at the heart of these innovations. As consumers increasingly turn to voice assistants and image-based searches to find what they need, AI enhances these search modalities to deliver more accurate and relevant results. For voice search, AI works by understanding and processing natural language, allowing it to comprehend and respond to user queries rapidly. This capability enables brands to optimize their content for voice search, ensuring they remain visible in this rapidly growing search domain. In visual search, AI analyzes images uploaded by users to identify products and similar items, offering an immediate link to purchase. This seamless integration of AI in search technologies not only enhances the user experience but also bridges the gap between visual inspiration and transaction, creating new pathways for consumer engagement that are intuitive and immediate.

As AI continues to redefine the boundaries of what is possible in marketing, it ushers in an era where personalization, real-time adaptability, and innovative search modalities converge to create marketing strategies that are effective and intimately aligned with consumers' individual preferences and behaviors. This shift marks a significant evolution in how brands connect with their audiences, moving from a one-size-fits-all approach to one that values and responds to the nuances of individual consumer journeys.

AI for Small Businesses: Cost-Effective Solutions

Small businesses, often operating with limited budgets and resources, need help to remain competitive and efficient. The advent of artificial intelligence (AI) in various business operations has opened a gateway for small businesses to leverage advanced technologies that were once the exclusive domain of more giant corporations. By integrating AI into their systems, small businesses can streamline their operations and enhance their ability to make data-driven decisions without incurring prohibitive costs.

Affordable CRM Systems

Customer relationship management (CRM) systems are fundamental for managing interactions with current and potential customers, a critical aspect of any business striving for growth and customer satisfaction. Traditionally, robust CRM systems have been costly, often requiring significant upfront and ongoing investment that small businesses could not afford. However, AI-powered CRM systems have changed the landscape by offering cost-effective solutions that provide the dual benefits of automation and deep customer insights. These AI-driven systems can automatically update records, track customer interaction histories, and predict future customer behaviors based on past data. This level of automation reduces the need for manual input, cutting labor costs and minimizing human errors. For a small business, this means maintaining a sophisticated CRM system that enhances customer relationships and drives sales while operating within a modest budget. Additionally, AI in CRM helps segment customers based on their behavior and preferences, allowing small businesses to tailor their marketing and sales strategies more effectively without extensive market research.

Automated Inventory Management

Inventory management can be a daunting challenge, especially for small businesses that need help to afford the losses associated with overstocking or the missed opportunities due to stockouts. AI has become a strategic tool that can transform this aspect of business by leveraging predictive analytics to optimize inventory levels. AI algorithms can accurately forecast demand by analyzing sales data, seasonal trends, and broader market conditions. This predictive capability enables small business owners to make informed decisions about stock levels, ensuring they have enough inventory to meet customer demand without spending too much capital on unsold goods. Moreover, AI can identify trends and patterns that may take time to be evident through manual analysis, such as subtle increases in product demand during certain times of the week or variations due to local events. By automating these analyses, AI helps maintain efficient inventory levels and reduces the time and effort spent on manual stock management, allowing business owners to focus on other critical areas of their business.

AI for Financial Planning

Financial planning is crucial for the survival and growth of any small business, but it requires a level of precision and foresight that can be challenging to maintain consistently. AI tools in financial planning come as a boon for small businesses, offering insights derived from comprehensive data analysis. These tools can precisely examine cash flow patterns, expenses, and revenues to provide forecasts and financial advice. By doing so, AI enables small business owners to anticipate financial needs and challenges, plan for future investments, and manage cash flow more effectively. For instance, an AI system might analyze transaction data to identify periods of cash flow shortages and suggest appropriate corrective actions, such as adjusting payment terms with suppliers or scheduling big-ticket purchases more strategically. This proactive approach to financial management helps small businesses avoid common pitfalls like undercapitalization or cash crunches, ensuring they remain financially healthy and free to pursue growth opportunities.

Accessible Marketing Insights

In the realm of marketing, knowledge is power. For small businesses, gaining access to sophisticated marketing analytics has historically been a challenge, often requiring either significant investment in market research or a trial-and-error approach that can be costly and inefficient. AI levels the playing field by providing small businesses access to powerful marketing insights from extensive data analysis. AI tools can track and analyze online consumer behavior, social media trends, and competitor activities, giving small businesses a detailed view of the market landscape. This information allows small businesses to craft targeted marketing strategies that are more likely to resonate with their audience, optimize marketing spend, and increase the overall effectiveness of their campaigns. For example, AI can help a small boutique identify which products are trending on social media and adjust its inventory and marketing efforts accordingly, capturing sales that might otherwise go to larger competitors with more established marketing infrastructures.

Integrating AI into small business operations democratizes access to technologies that enhance efficiency, reduce costs, and improve decision-making. By adopting AI solutions, small businesses level the playing field and position themselves to compete more effectively in an increasingly digital marketplace, ensuring their longevity and success in the face of challenges.

The Future of AI in Remote Work

As the work landscape undergoes a seismic shift toward remote setups, AI emerges as a pivotal force in redefining how teams collaborate, secure their data, learn, and balance their professional and personal lives. The transformation driven by AI in remote work environments highlights its role as a technological tool and as a facilitator of more dynamic, responsive, and efficient work models.

Remote Collaboration Tools

In remote work, seamless collaboration is beneficial and essential for maintaining productivity and cohesion among distributed teams. AI enhances remote collaboration tools by streamlining communication and project coordination, ensuring that geographical distances do not translate into operational inefficiencies. For instance, AI-integrated project management software can automatically organize and prioritize tasks based on deadlines, team member availability, and project status. This capability allows team leaders to focus more on strategic decision-making rather than getting bogged down by the minutiae of project coordination. Furthermore, AI-driven collaborative platforms can offer real-time language translation, making it easier for global teams to interact without language barriers, thus fostering a more inclusive work environment. By enhancing these tools, AI supports the logistical aspects of remote work and improves the social dynamics, ensuring that team members feel connected and engaged regardless of their physical locations.

AI-Driven Cybersecurity for Remote Work

As remote work becomes more prevalent, so does the complexity of maintaining robust cybersecurity measures. AI significantly bolsters cybersecurity in remote environments by monitoring network activities and identifying potential threats in real time. Through continuous learning algorithms, AI systems can detect anomalies that deviate from standard usage patterns, such as unusual login attempts or suspicious data transfers. Upon detecting these threats, AI can initiate automatic countermeasures, such as alerting administrators, blocking malicious activities, or even isolating affected systems to prevent the spread of potential breaches. This proactive approach to cybersecurity protects sensitive information. It ensures that the integrity of remote work environments is maintained, allowing team members to focus on their tasks with peace of mind about their digital safety.

Personalized Learning Platforms

Continual learning and adaptation are essential in today's rapidly changing professional landscapes, and AI plays a crucial role in facilitating these processes for remote workers. AI-powered learning platforms tailor educational content to meet the specific needs of individuals, adapting to their learning pace and preferred content format and even identifying knowledge gaps that need addressing. These platforms can suggest personalized learning paths, including online courses, workshops, and seminars most relevant to the user's career goals and current job requirements. For example, an AI learning platform might analyze users' project management tasks and recommend specific leadership courses to enhance their skills. This personalized approach makes learning more effective and engaging, as workers feel that their professional development aligns with their personal aspirations and job requirements.

Work-Life Balance Optimization

The line between professional and personal life can often blur in remote work settings, making work-life balance a critical concern. AI helps manage this balance by analyzing work patterns and suggesting optimal schedules and breaks. For instance, AI tools can identify when a worker is most productive and suggest focusing on intensive tasks while reserving other times for meetings or administrative tasks. Additionally, AI can prompt users to take breaks at optimal times to prevent burnout—perhaps after long periods of continuous work or when patterns suggest diminishing productivity. These suggestions help individuals maintain a healthy work-life balance, contributing to their well-being and productivity.

In summary, AI's integration into the fabric of remote work transforms how teams collaborate, how security is managed, how individuals learn and grow professionally, and how they balance work with personal life. As we forge ahead, the role of AI in facilitating these aspects will undoubtedly expand, making remote work a necessity in times of crisis and a viable, efficient, and enriching mode of professional engagement. In the next chapter, we will explore how AI reshapes the creative industries, turns traditional practices on its head, and opens up new possibilities for innovation and expression.

Show Others How AI Can Unlock Unimaginable Productivity and Innovation

"Artificial intelligence, deep learning, machine learning — whatever you're doing if you don't understand it — learn it. Because otherwise, you're going to be a dinosaur within three years."

Mark Cuban

The words "AI" probably conjure many images in people's minds—from enhanced work productivity to robots creating ever-refined versions of themselves. In your day-to-day life, however, when discussing AI tools, the most oft-mentioned tool is ChatGPT. Sure, this handy tool makes so many aspects of our working life much smoother and more dynamic than in the past... But by this stage of your reading journey, you have seen that AI is pertinent in almost all industries. From helping teachers create personalized lesson plans for their students to inspiring artists to create avant-garde pieces, it has so much scope, and the pace of its progress is unstoppable.

As human beings, it is vital to understand the magnitude and ethical implications of AI. We encounter AI daily through our voice assistants, curated social media feeds, e-commerce chatbots, and Google Maps. And this is just scratching the surface. We can obtain so much more from AI by knowing and testing its capabilities.

Thus far, we have focused on how AI can enhance your everyday life and the lives of students and workers in various professions and tasks—including customer service, marketing, financial planning, and communication for remote workers. If what you see inspires you to delve deeper into AI, I hope you can share your opinion of this book.

By leaving a review of this book on Amazon, you'll let new readers know where they can find a book that enlightens them on the surprising ways AI can enhance their productivity, creativity, and enjoyment of life.

Thanks for your support. The best is yet to come... keep reading to discover how AI can aid you in your journey to become a healthier, happier, more mindful person!

SCAN ME

FIVE

AI and the Arts—A New Creative Landscape

As the sun sets on traditional artistic methods and the dawn of digital innovation rises, the creative landscape is being redefined by a silent yet profound revolution powered by artificial intelligence (AI). Imagine a world where the music moves and evolves with you, where compositions adapt to your emotions and preferences in real-time. This is not the distant future; this is the now, where AI serves as a dynamic to composers and musicians, offering tools that extend their creative capabilities and redefine the boundaries of musical expression. In this chapter, we explore how AI is a tool for creation and a partner in performance, enhancing the auditory experience for creators and audiences alike.

AI in Music Production: A Composer's Ally

Enhancing Composition

AI's ability to analyze and learn from patterns has found a particularly harmonious application in music production. AI tools, equipped with algorithms that understand music theory, can now assist composers by suggesting chord progressions and harmonizations that enhance their musical pieces. These suggestions are not random but are based on an analysis of vast amounts of music data, from classical compositions to contemporary hits, allowing AI to understand trends and user preferences. This capability is particularly invaluable for emerging artists and educators, who can explore complex musical styles and techniques without years of specialized study. For instance, a music teacher can use AI software to demonstrate to students how different arrangements affect the mood and style of a piece, making lessons both engaging and informative. By providing these AI-driven insights, the barrier to creating sophisticated and appealing music is significantly lowered, democratizing music production and allowing a broader range of artists to express their unique creative visions.

Sound Design and Synthesis

Beyond traditional composition, AI is pushing the frontiers of sound design and synthesis. AI systems can generate or alter new sounds through machine learning algorithms, creating previously impossible or unimaginable auditory experiences. This technology taps into a deep neural network's ability to mimic the vast complexities of sound, allowing it to produce rich, organic textures or entirely new sounds that bring a fresh dimension to music production. Artists and technologists can experiment with these sounds, incorporating them into performances or installations to create a more immersive experience for their audience. The flexibility of AI in sound design not only expands the palette of available sounds but also inspires new forms of musical expression, challenging artists to explore the uncharted territories of their creative landscapes.

Real-Time Performance Collaboration

One of AI music technology's most exhilarating advancements is its ability to perform live in collaboration with human musicians. AI algorithms can now respond in real-time to music played during live performances, adapting to tempo, style, and harmony changes. This interaction creates a dynamic performance environment where human creativity and AI responsiveness blend to produce a unique and unpredictable auditory experience. For performers, this means that AI becomes a virtual band member, capable of accompanying them with sensitivity and adaptability traditionally associated only with human musicians. This technology enhances live performances and offers new ways to teach dynamic composition and improvisation in educational settings, providing students with an immediate and responsive music-making experience.

Music Personalization for Listeners

The influence of AI in music extends beyond creation and performance to how listeners experience music. AI-driven platforms can now personalize music streaming services according to the listener's mood, activity, or preferences. This personalization is achieved by analyzing user data and behavioral patterns to predict what music might suit the listener at a particular moment. Whether energizing tracks for a workout session or calming melodies for a night in, AI ensures that the music aligns with the listener's current state, enhancing the overall listening experience. This tailored approach makes listening more enjoyable and deepens the listener's connection to the music, making it feel uniquely theirs.

As we continue to explore AI's role in the arts, it becomes evident that the technology's potential to transform creative processes is profound and multifaceted. From enhancing musical compositions to personalizing listener experiences, AI is not just a tool for artists but a collaborator that brings new dimensions to artistic expression. As AI technologies evolve and become more integrated into the arts, they promise to unlock new creative potentials, redefine artistic boundaries, and enrich the cultural landscape of our society. As we tune in to the next section, let's keep our minds open to the symphony of possibilities AI conducts in the ever-evolving art world.

Generative Art: AI as a Collaborative Artist

In the vibrant nexus where technology meets art, generative adversarial networks (GANs) and similar AI systems are revolutionizing the creative process, ushering in a new era where AI acts as a tool and co-creator. These sophisticated algorithms engage in digital dialogue, where one part of the system generates new images. At the same time, another evaluates them against a set of criteria, iterating toward an ever-evolving artistic ideal. This process mirrors the traditional method where an artist might sketch, critique, and refine their work. Still, it occurs at a pace and complexity that only digital processes can sustain.

Imagine walking into a gallery where the art pieces you view are created not only by these AI systems but are also unique to each visitor, crafted in real-time based on personal data inputs like emotional state or aesthetic preferences. Such is the potential of AI in art to create not just static images but dynamic interactions that evolve before your eyes. This adaptability makes AI an invaluable partner in the artistic process, providing artists with a dynamic palette not limited by human creators' physical and cognitive constraints. Moreover, these AI collaborators bring detachment and non-bias to art creation, potentially challenging human artists to explore new perspectives and push beyond their conventional boundaries.

Moving beyond the canvas, AI's role in interactive art installations illustrates how technology can transform passive viewing into an active and immersive experience. Through sensors and real-time data processing, AI-driven installations can respond to the presence and movements of viewers, changing and adapting the visual or auditory output to create a personalized experience. For example, an installation might alter its color scheme based on the viewer's movement speed or play different soundtracks in response to the number of people in the space. This level of interactivity invites the audience to become part of the art itself, blurring the lines between creator and observer and between art and experience.

AI's influence also extends into conceptual art, which challenges the notions of creativity and authorship. Artists are using AI to question and explore new concepts in art, pushing the boundaries of what is considered possible or acceptable in traditional mediums. AI algorithms can be programmed to follow, deviate from, or completely disregard established artistic rules, creating works that can be unpredictable and provocative. This capability allows artists to explore complex ideas in previously impossible ways, such as the intersection of technology and identity or the conflict between nature and digital realities. By integrating AI into their practice, artists can delve into these themes with depth and authenticity that resonate with contemporary audiences, making abstract concepts tangible and relatable.

Moreover, AI's ability to analyze and learn from historical art styles offers a fascinating avenue for preserving and evolving artistic traditions. By feeding AI systems with vast databases of historical art, artists can generate works that reflect the influences of past styles while incorporating contemporary elements. Combining old and new helps preserve cultural heritage and ensures its relevance in the modern world. For instance, an AI could be trained in the techniques of Renaissance painters, enabling it to produce artworks that echo the complexity and depth of that era but with modern themes and messages. This blending of times and styles enriches the artistic landscape. It provides a meaningful link between past and present, inviting reflection on how art evolves and reflects human history and society.

In this exploration of AI as a collaborative artist, a portrait of potential emerges—a landscape of artistic expression that is richer and more diverse because of AI integration. As these technologies continue to develop, their potential to transform the arts remains boundless, promising new ways for artists to explore, create, and engage with their audiences.

AI in Literature: Penning the Next Bestseller

In the quietly bustling world of literature, where the clatter of type-writer keys has given way to the soft clicks of computer keyboards, AI is emerging as a significant ally in the creative process. Imagine sitting down to write, grappling with the vast possibilities of a blank page, and finding a partner in an AI tool that helps you navigate the maze of narrative construction. AI in literature is transforming the solitude of writing into a collaborative endeavor with technology, offering tools that spark creativity and streamline the intricate storytelling process.

Narrative Generation

For authors, the genesis of a story often begins with a single idea, a fleeting thought that needs nurturing to blossom into a complex narrative. AI is a brainstorming partner, using its vast database of literature and narrative structures to suggest plot twists, character developments, and thematic elements. AI tools employ natural language processing to generate narrative ideas that writers can use as story springboards. These tools analyze thousands of narratives to identify patterns and elements that resonate with readers, making suggestions that are not just mechanically sound but emotionally engaging. For instance, if you're writing a mystery novel, AI can propose scenarios that introduce red herrings or unexpected revelations based on successful narrative structures from best-selling mystery books. This support is invaluable, especially for novice writers or educators in creative writing, as it democratizes the ability to craft compelling stories by providing insights typically gained through years of experience and study.

Language Modeling and Text Editing

Beyond the generation of ideas, AI profoundly impacts the refinement of language in literary works. AI-powered editing tools do more than correct grammar and spelling; they enhance style, coherence, and text's overall readability. These tools analyze the context of written content, suggesting changes that improve flow and clarity while maintaining the author's unique voice. For example, if a text is overly complex or jargon-heavy, AI can suggest simpler alternatives that make the content accessible without diluting its meaning. This capability makes editing more efficient and helps writers focus more on the creative aspects of writing. They can experiment with different narrative styles or delve deeper into character development, knowing AI can effectively manage linguistic polish. This shift accelerates the editing process and elevates the quality of writing, making literature a more inclusive and expressive art form.

Translation and Localization

The power of literature lies in its ability to transcend boundaries, and AI plays a pivotal role in extending the reach of literary works across the globe. AI-driven translation tools can preserve the nuances and subtleties of the original text, ensuring that translations do not become mere word-for-word transpositions but maintain the spirit and style of the source material. These tools use advanced algorithms to understand idiomatic expressions, cultural references, and stylistic nuances, adapting them appropriately for different languages and regions. This precision not only broadens the audience for authors but also preserves the cultural richness of literature, allowing stories to resonate with readers from diverse backgrounds. Furthermore, AI in localization goes beyond text, adapting book covers, marketing materials, and promotional strategies to fit cultural preferences and norms, ensuring that literature is seen and felt in its translated forms.

Customized Reading Experiences

AI's impact on literature is not confined to the realm of creation but extends to how books are consumed. AI technologies are now used to tailor reading experiences to individual preferences, adjusting the complexity, style, and even the pacing of narratives based on a reader's past interactions and feedback. This personalization transforms reading from a static to a dynamic experience, where the narrative can evolve in real-time to match the reader's mood or changing preferences. For instance, a reading app powered by AI might offer alternate plot developments in a novel, allowing readers to choose paths that interest them, much like a choose-your-own-adventure book but with a sophisticated algorithmic twist. This level of customization makes reading more engaging and immersive, as readers see their reflections in the stories they explore, fostering a deeper connection with the literature.

AI continues to weave its threads through the tapestry of literary creation and consumption, redefining the boundaries between author and audience, creation and interaction. In this evolving landscape, AI is not just a tool but a catalyst for innovation, opening new chapters in the story of literature where the possibilities are as limitless as the imagination.

Ethical Considerations in AI-generated Art

As artificial intelligence (AI) continues to permeate the creative realms, it offers vast possibilities and ushers in a series of ethical dilemmas, particularly around authorship and copyright, authenticity and originality, bias in representation, and the economic impacts on human artists. These issues not only challenge our understanding of creation and creativity but also mandate a critical evaluation of the implications that AI-generated art holds for the broader cultural and legal landscape.

Authorship and Copyright

One of the most pressing concerns in the intersection of AI and art revolves around the issues of authorship. When a piece of art is created with significant AI involvement, determining the copyright holder becomes complex. Traditionally, copyright laws are designed to protect the interests of human creators, but AI disrupts this understanding by introducing nonhuman agency in the creative process. If an AI algorithm trained on thousands of existing artworks generates a new piece, the question arises: Who is the creator? Is it the developer who designed the AI, the user who inputted the specific parameters, or the AI itself? And, more importantly, who holds the copyright to such works? These theoretical questions have practical implications for producing, sharing, and monetizing art. Resolving these issues might require new legal frameworks that recognize the unique contributions of AI systems while ensuring that human creators are fairly compensated for their creative inputs.

Authenticity and Originality

Furthermore, the integration of AI in art raises profound questions about authenticity and originality—concepts that are cornerstones of the value and appreciation of artistic works. AI-generated art, by its nature, challenges these notions as it often involves replicating and recombining elements from existing works. This capability of AI to mimic and remix has led to debates over whether such works can genuinely be considered original or authentic. The impact of this on the perception of art is significant. It compels artists and audiences to reconsider what constitutes originality in the digital age. Does the value of art lie in its novelty, or can the reinterpretation of familiar themes and styles through AI offer a new form of authenticity? Addressing these questions is vital for understanding AI's role in art and maintaining the integrity and appreciation of art in society.

Bias and Representation

Moreover, the algorithms that drive AI are not free from biases, as they learn from datasets that may reflect historical biases or cultural stereotypes. This can result in AI-generated art that inadvertently perpetuates these biases, influencing which narratives are told and whose images are celebrated or marginalized. For instance, an AI system trained predominantly in Western art styles might fail to represent accurately or even misinterpret themes from non-Western cultures. This not only skews representation but also impacts cultural diversity in art. Ensuring AI systems are trained on diverse and inclusive datasets is crucial to mitigating these biases. Artists and technologists need to collaborate to develop AI that upholds the principles of diversity and equity, ensuring that the art it produces is innovative and inclusive.

Economic Impacts on Artists

Finally, the rise of AI in creating and selling art presents opportunities and challenges for human artists. On the one hand, AI can democratize art production, enabling more individuals to make and sell art by lowering barriers to entry and reducing the need for traditional art education and skills. On the other hand, it could also lead to economic displacement for those artists whose livelihoods depend on selling traditional, human-made art. Introducing AI-generated artwork in the market could drive down the value of human-created art or make it more difficult for artists to differentiate themselves. Navigating these economic impacts requires a balanced approach that leverages AI's capabilities to expand the art market and enhance artistic creativity without undermining the financial sustainability of human artists.

In navigating these ethical considerations, stakeholders from across the art world—creators, curators, collectors, and legal experts—must engage in ongoing dialogue to address the challenges posed by AI-generated art. By doing so, they will not only shape the trajectory of AI in the arts but also ensure that this evolution reflects our collective values and respects the rights and roles of all individuals involved.

Promoting Cultural Appreciation Through AI in Art

In an era where digital technologies redefine boundaries, artificial intelligence (AI) emerges as a formidable catalyst in the fusion of diverse cultural expressions. This transformative tool facilitates unique collaborations among artists from various cultural backgrounds, creating a vibrant tapestry that celebrates a global heritage. Imagine a digital workspace where artists from Tokyo, Lagos, and São Paulo simultaneously contribute to a single canvas, each infusing it with distinct cultural symbols and narratives. AI algorithms manage the logistical aspects of such collaborations in this setting and intelligently analyze and blend different artistic elements to ensure coherence and harmony in the final piece. This seamless integration encourages a deeper understanding and appreciation of diverse cultural expressions, fostering a spirit of global artistic unity.

Moreover, AI's capability extends to preserving and revitalizing cultural heritage through the digital restoration and preservation of artworks. Tools powered by AI meticulously analyze the condition of artworks, predict potential deterioration, and recommend the best conservation methods. This helps maintain the physical integrity of artworks and ensures that the cultural narratives embodied in these pieces are passed down to future generations. For instance, AI applications can digitally reconstruct damaged parts of ancient frescoes, allowing us to view them as they once were without physically altering the original works. This preservation is crucial for educational purposes, providing scholars and students access to historical artworks in unprecedented detail, enhancing academic studies, and fostering a deeper appreciation of cultural history.

Within the walls of museums, AI-driven tools are revolutionizing how visitors engage with art. By customizing educational content based on a visitor's past interactions, preferences, or emotional responses, AI creates an informative and engaging personalized museum experience. For example, a visitor interested in Gothic architecture might be guided to a virtual tour highlighting critical European Gothic structures, complete with AI-generated explanations and comparisons. This tailored approach enriches the visitor's experience and enhances their understanding of art in a broader cultural context, making their visit more meaningful and impactful.

Expanding the horizon further, AI-powered global art platforms are democratizing access to art by breaking down geographical and economic barriers. These platforms utilize AI to curate and recommend artworks from diverse cultures based on users' interests and previous interactions. Such exposure broadens the audience's artistic horizons and elevates underrepresented artists to a global stage, ensuring local traditions and modern expressions find their place in the international art scene. This global connectivity fosters a greater appreciation for the richness and diversity of world art traditions, promoting cultural understanding and appreciation on a scale never seen before.

In weaving together the threads of technology, art, and culture, AI acts as a bridge connecting diverse artistic landscapes across the globe. Through collaboration, preservation, education, and exposure, AI enriches our cultural experiences, allowing us to appreciate the vast tapestry of global heritage in more profound and personal ways. As we close this chapter on AI's role in promoting cultural appreciation through art, we look forward to exploring further how this technology continues to shape our understanding and interaction with the world around us. In the next chapter, we delve into how AI influences another pivotal area of human experience: personal development and lifestyle, where it promises to reshape habits, enhance learning, and enrich daily living in ways we are only beginning to understand.

SIX

AI for Personal Development and Lifestyle

I n a world where personal development is increasingly intertwined with digital innovation, artificial intelligence (AI) emerges as a transformative force in sculpting a lifestyle that is not only healthier but also more attuned to individual needs and aspirations. Picture a scenario where your fitness coach is always available and perpetually attuned to your physical condition, progress, and goals. This chapter delves into how AI, particularly in fitness and well-being, is revolutionizing how we approach personal health, offering custom-tailored guidance that adapts to our evolving needs in real-time.

AI Fitness Coaches: Personalized Training Programs

Adaptive Workout Plans

One of the most dynamic aspects of AI in fitness is its capability to craft personalized workout plans that adapt in real-time, ensuring each session is optimized for maximum benefit. AI fitness coaches process data from past workouts to predict and shape future sessions, making adjustments based on your progress and any plateau in development. For example, suppose your goal is to increase stamina. In that case, AI algorithms analyze your pace, heart rate, and fatigue levels during each session and adjust the intensity or duration of your workouts accordingly. This dynamic planning helps maintain an optimal challenge, crucial for continuous improvement, while also adapting to your body's needs, reducing the risk of overtraining or injury.

Injury Prevention

AI's predictive capabilities are particularly impactful in preventing injuries. By analyzing data on your movements and comparing them to optimal workout forms, AI can identify potential risk patterns that may lead to injury. For instance, if an AI system notices that your form deteriorates dramatically at specific points in your run—an indicator of potential knee stress—it can suggest corrective exercises to strengthen related muscles or recommend a decrease in intensity to allow recovery. This proactive approach keeps you safe and ensures that preventable injuries do not derail your fitness journey.

Integration with Wearables

The synergy between AI and wearable technology multiplies the benefits of digital fitness tools. Wearables collect vital data such as heart rate, calories burned, and activity levels, which AI systems analyze to provide insights into your health and fitness. This integration allows for a holistic view of your wellness that goes beyond simple activity tracking. For example, by monitoring your heart rate variability, AI can assess your stress levels and recovery status, then adjust your workout and rest days accordingly. This comprehensive health management ensures that every aspect of your fitness—from exercise intensity to recovery—is optimized for your needs.

Motivation and Engagement

One of the most crucial roles of AI in fitness is its ability to keep you motivated and engaged. AI fitness coaches are designed not just to guide but also to inspire. They achieve this by understanding your motivation triggers and learning from your interactions. For example, if you are goal-oriented, AI will continuously remind you of your progress and celebrate milestones to keep you motivated. If you thrive on variety, the AI can introduce new exercises to keep the routine exciting. Furthermore, AI can create personalized challenges and set realistic yet ambitious goals based on your performance, making the fitness journey both rewarding and enjoyable.

As we explore AI's expansive roles in personal development and lifestyle, it becomes clear that AI's influence goes beyond mere convenience—it becomes a pivotal partner in our quest for a healthier, more balanced life. Through its integration into fitness, AI enhances physical well-being and empowers us to take proactive control of our health, transforming the landscape of personal fitness and wellness with every step we take.

Nutritional AI: Tailoring Diet Plans to Genetic Profiles

In the quest for optimal health, the adage "one size fits all" no longer applies, especially regarding nutrition. Enter the realm of artificial intelligence (AI), paving the way for revolutionary changes in how we approach our diets. By harnessing the power of AI to analyze genetic data, personalized nutrition plans are becoming not just a possibility but a practical reality. This novel approach considers individual genetic differences that impact metabolism, nutrient absorption, and even food sensitivities, offering a tailored diet strategy that significantly enhances personal health and wellness.

AI's role in transforming dietary planning starts with its ability to process vast amounts of genetic information. By examining specific biomarkers in your DNA, AI algorithms can identify metabolic rates and genetic predispositions related to nutrient processing. For instance, some individuals might have a genetic variant that affects how they metabolize caffeine or lactose. With this information, AI can develop personalized diet plans that emphasize or limit certain foods, aligning perfectly with your body's unique needs. This hyper-personalization optimizes your nutritional intake and aids in preventing diet-related diseases such as diabetes or heart disease by recommending preventive diets.

Moreover, AI extends its capabilities to accommodate individual allergies and dietary restrictions seamlessly into meal planning. If you have a known allergy to nuts or a gluten intolerance, AI systems can automatically filter out recipes or food items from your diet plan containing these allergens. This feature is particularly beneficial for those who might find tracking such details cumbersome or those at risk of severe allergic reactions. The AI ensures safety and convenience, allowing you to enjoy various meals without worrying about potential allergic reactions or dietary compliance.

Real-Time Dietary Tracking

Moving beyond static meal plans, AI also revolutionizes how we track and manage our dietary habits. Real-time dietary tracking through AI involves sophisticated image recognition technology that can analyze photos of your meals. These AI tools assess portion sizes, cooking methods, and ingredients to provide immediate nutritional feedback. For example, by simply taking a photo of your lunch, AI can calculate calorie intake and macronutrient distributions and suggest modifications to improve meal balance. This immediate feedback loop helps maintain dietary discipline and educates you on making healthier food choices independently. Over time, this continuous interaction with AI dietary tracking encourages a deeper understanding of nutrition and its impact on your body, fostering a more mindful approach to eating.

Sustainable Eating Suggestions

In today's environmentally conscious world, AI also plays a crucial role in promoting sustainable eating habits. AI can suggest meals that minimize your carbon footprint by analyzing local seasonal produce and available sustainable food sources. For instance, if you're in a coastal area, AI might prioritize seafood from sustainable fisheries or locally grown vegetables. These suggestions help reduce the environmental impact associated with long-distance food transportation and support local economies. Additionally, AI can provide insights into food waste management by suggesting recipes that utilize leftovers effectively, promoting a zero-waste lifestyle that is both healthy and sustainable.

Through these diverse applications, AI sets a new standard in nutritional planning and management, making it a pivotal tool in pursuing a healthier lifestyle tailored to individual needs. As we explore AI's potential in personal development, it becomes clear that its impact on our dietary habits is profound, offering a more thoughtful, safer, and more sustainable way to eat.

Mindfulness and Meditation: AI-Driven Wellness Tools

In the contemporary era, where stress often accompanies the bustling rhythm of daily life, the ancient practices of mindfulness and meditation have found a new ally in artificial intelligence (AI). Embracing the serenity that meditation offers no longer requires solitude in scenic retreats; AI brings this tranquility directly to your personal space, explicitly tailored to your emotional and psychological needs. AI-driven applications in mindfulness and meditation are revolutionizing how individuals engage with mental wellness practices, making them more accessible, personalized, and effective than ever before.

Customized meditation sessions are at the forefront of this revolution. AI technologies integrated into apps and smart devices can now analyze data regarding your current mood and stress levels, collected through user input or physiological sensors. For instance, suppose you report feeling particularly stressed after a long day at work; the AI can suggest a specific meditation sequence that targets stress relief, perhaps focusing on deep breathing techniques or guided imagery that has proven effective for you in the past. The customization doesn't end here. Over time, the AI learns which techniques yield the best outcomes for you, continuously refining its recommendations to better align with your evolving preferences and emotional states. This personalized approach ensures that each meditation session is optimally beneficial, helping you achieve excellent balance and peace in your daily life.

Biofeedback Integration

Integrating biofeedback mechanisms with AI significantly enhances the effectiveness of mindfulness exercises by providing real-time physiological data that guide your practices. Devices equipped with sensors to monitor body signals such as heart rate, skin conductance, and even brain waves offer a window into your body's responses to different meditation techniques. AI algorithms analyze this data to identify the most effective strategies for reducing your stress or anxiety levels. For example, suppose the AI detects that your heart rate does not decrease during a relaxing meditation. In that case, it might suggest switching to a different style, perhaps from passive meditation to an active one involving mindful movement, to better suit your body's needs. This tailored guidance helps cultivate a genuinely effective personal meditation practice that adapts to your body's unique responses, fostering enhanced mental wellness and self-awareness.

Progress Tracking and Analysis

AI's capability to track and analyze your progress in mental wellness practices is invaluable in a journey toward sustained mental health. These AI systems monitor how frequently you engage in mindfulness exercises and measure the tangible benefits these practices have on your well-being. AI provides insightful feedback by charting your stress levels, mood improvements, and even sleep patterns over time, highlighting your progress and pinpointing areas needing additional focus. For instance, if the AI notices a correlation between increased meditation frequency and improved sleep quality, it might encourage more consistent practice times, reinforcing habits that positively impact your sleep. This ongoing analysis motivates you by showing tangible results and personalizes the journey, making adjustments based on what works best for your lifestyle and mental health needs.

Virtual Wellness Coaches

Imagine having a wellness coach who is always available and provides immediate support and guidance whenever needed—this is what AI-powered virtual coaches offer. These digital mentors use AI to deliver personalized advice and support, mimicking your interactions with a human coach. Through conversational AI, these virtual coaches can respond to your queries about mindfulness techniques, offer encouragement, or provide reminders to practice regularly. They can even lead you through meditation sessions, adjusting the pace and focus of the session in response to your real-time feedback. This constant availability makes maintaining regular mindfulness practices easier. It ensures that your support is consistent and tailored to your evolving needs, making mental wellness an integrated part of your daily routine.

Through these personalized, insightful, and accessible applications, AI sets a new standard in how we approach mindfulness and meditation. By adapting to individual needs and providing continuous support, AI-driven wellness tools are not just enhancing our meditation experiences; they are transforming our overall approach to mental health, making it more proactive, personalized, and attuned to the complexities of modern life. As we continue to explore and integrate these technologies, they promise to play a pivotal role in shaping a healthier, more mindful society.

AI and Hobbies: Enhancing Personal Pursuits

In the fabric of daily life, where routines often dictate the rhythm, hobbies serve as a personal escape—a chance to cultivate joy, creativity, and skill in the corners of time we call our own. Artificial intelligence (AI) is stepping into these spaces, augmenting the way hobbies are learned, practiced, and enjoyed, making these personal pursuits not only more accessible but also more enriching. AI's role in this aspect of life underscores its versatility and capacity to enhance human experiences beyond productivity and professional demands, venturing into personal satisfaction and leisure.

AI platforms are revolutionizing the way individuals engage with and learn new hobbies. These platforms utilize sophisticated algorithms to offer customized tutorials and adaptive learning paths explicitly tailored to an individual's pace of learning and areas of interest. For instance, someone keen on learning to play the guitar can use an AI-driven music learning app that listens to their playing in real-time, provides immediate feedback on timing and accuracy, and adjusts lesson plans based on their progress. This personalized feedback loop accelerates the learning process. It keeps engaging by ensuring the challenges are well-suited to the learner's current skills, pushing just enough to stretch their capabilities without causing frustration or disinterest. The technology behind this, often incorporating video analysis and machine learning, allows the AI to observe minute details in the user's actions, from finger placement to strumming technique, offering specific guidance that might be difficult for human instructors to monitor consistently.

Turning to art and creativity, AI tools are opening new avenues for artists and creative individuals by providing them with innovative techniques and suggestions that push the boundaries of traditional art forms. Digital artists, for example, can use AI-powered software to experiment with different styles and techniques, such as digital brush strokes or color theory, which the AI has learned from analyzing thousands of artworks. The software can suggest modifications to a piece that enhance visual appeal or convey emotion more powerfully based on learned patterns of human emotional responses to different colors and forms. Moreover, AI applications can generate base tracks for music composition that serve as the foundation for new music pieces, allowing musicians to modify and build upon them with their unique inputs. This streamlines the creative process and inspires artists by blending their creativity with algorithmic innovations, leading to unique compositions that might not have been conceived without AI's input.

In gardening and home improvement, AI applications are proving invaluable by offering advice tailored to the specific conditions of a user's environment and skill level. For gardening enthusiasts, AI-driven gardening apps analyze soil type, climate data, and plant health through sensors or user input to provide customized gardening advice. These apps can alert you about the best planting times for each type of seed in your location or offer solutions to plant diseases identified via uploaded photos of the affected plant. Similarly, AI in home improvement can guide users through DIY projects by considering the user's proficiency and the specifics of their home environment. For instance, an AI home improvement helper might suggest a list of tools and materials based on the user's past projects and offer step-by-step customization instructions that fit the user's skill level and the dimensions of their living space.

Interactive gaming experiences, too, are being transformed by AI, which enhances gameplay by adapting to the individual player's style and preferences. In video games, AI algorithms can modify game difficulty in real-time, responding to how well the player does to provide a consistently challenging experience. More advanced AI systems can even alter storylines based on players' decisions, creating a profoundly personalized narrative that can lead to multiple endings. This not only makes gaming more engaging but also increases the replay value of games, as each playthrough can offer a new experience based on different choices made by the player. AI's role in gaming demonstrates its potential to create highly immersive and responsive entertainment experiences that adapt to and reflect the player's decisions and style, uniquely satisfying each gaming session.

As AI continues to permeate our spheres, enhancing how we engage with our hobbies and leisure activities, it becomes clear that AI's potential to enrich human life knows no bounds. From learning new skills to diving deeper into creative endeavors, from nurturing a garden to reshaping interactive gaming, AI stands out not merely as a tool of efficiency but as a companion in our quests for growth and enjoyment.

AI in Personal Safety Devices

In an age where safety and security are paramount, artificial intelligence (AI) is revolutionizing how we protect ourselves and our loved ones. By leveraging AI, personal safety devices have become more proactive, reacting to emergencies, anticipating potential threats, and responding appropriately. These advanced systems enhance our ability to maintain safety in an increasingly unpredictable world.

Emergency Response Systems

Imagine you or a loved one experiencing a fall at home or facing a sudden health emergency. AI-enhanced personal safety devices with sensors and intelligent algorithms can detect these unusual situations immediately. Upon detection, they don't just emit local alarms; they go further by automatically contacting emergency services, providing them with precise location data and details about the situation. This swift action ensures that help is quickly coming, which can be crucial in preventing severe outcomes. The sophistication of AI allows these systems to differentiate between everyday activities and genuine emergencies, minimizing false alarms and ensuring that help is summoned only when truly needed. This capability brings peace of mind to users and enhances the efficiency of emergency response services, ensuring they are alerted and informed accurately and promptly.

Personal Security Apps

AI's role in personal security extends into the digital realm, where personal security apps scrutinize your surroundings through real-time data analysis. These apps utilize AI to assess the safety of different areas based on vast amounts of data, including crime rates, recent incidents, and user-generated reports. If you plan to visit a new city or even a new area in your hometown, these apps can alert you about potentially unsafe areas or suggest safer routes. The real-time processing capabilities of AI mean that you receive timely notifications about any changes in the security status of your locations, allowing you to make informed decisions about where and when to visit. This proactive approach to personal security empowers you to avoid potential threats before they materialize, ensuring your safety as you navigate different environments.

Child and Elder Monitoring

For those caring for children or elderly family members, AI-powered devices offer a new level of vigilance and peace of mind. These devices use AI to monitor movements and activities. They send alerts if anything unusual is detected, such as a child straying into a restricted area or an elderly family member not following their usual daily routine. The sophistication of AI allows these systems to learn from daily patterns, making their monitoring more accurate and less intrusive over time. For caregivers, this means being able to keep an eye on their loved ones even when they're not physically present, ensuring their safety and well-being at all times. Additionally, these devices can monitor health-related metrics, alerting caregivers to potential health issues that may require immediate attention, such as sudden changes in heart rate or activity levels.

Data Privacy and Security

While AI brings numerous benefits to personal safety devices, it raises significant concerns about data privacy and security. These devices collect and process vast amounts of sensitive personal information, making them potential targets for cyber threats. Maintaining stringent data privacy and security measures is crucial to protect this information from unauthorized access and breaches. AI systems are designed with advanced security features, including encryption and anomaly detection algorithms, which identify and mitigate potential threats in real-time. Furthermore, developers and manufacturers must adhere to strict data protection regulations to handle user data responsibly. By prioritizing data security and privacy, AI-powered safety devices provide physical safety and safeguard their users' digital footprint, ensuring comprehensive protection in all aspects of their operation.

As we integrate these AI-driven safety tools into our daily lives, we enhance our ability to respond to emergencies and prevent them. The intelligent features of these devices make them indispensable tools in our quest for safety and security, reflecting AI's potential to transform our lives not just through convenience and efficiency but also through enhanced protection and peace of mind.

This exploration of AI's role in personal safety devices encapsulates its profound impact on our security and emergency management approach. By harnessing AI, we enhance our ability to protect ourselves and our loved ones, showcasing the technology's potential to change how we live and safeguard our lives. As we close this chapter, we look ahead to further discoveries in AI applications, each promising to broaden our understanding and integration of this transformative technology into every facet of our existence.

SEVEN

Addressing AI Misconceptions and Fears

I n the vast and ever-expanding universe of technology, artificial intelligence (AI) is a beacon of promise and a source of perennial myths. As you navigate through your daily digital interactions, from using your smartphone to browsing through streaming services, AI subtly enhances these experiences, often unnoticed. Yet, despite its pervasive presence, AI is frequently misunderstood, shrouded in a mist of sensationalism and speculative fiction. This chapter aims to clear the air, demystifying AI by separating fact from fiction and addressing the root causes of common misconceptions that obscure our understanding of this transformative technology.

Myth vs. Reality

The narrative surrounding AI is often peppered with dramatic assertions—robots taking over the world, AI surpassing human intelligence to become an omnipotent force, or AI as the panacea for all human challenges. These portrayals, while captivating, are far removed from the current capabilities and intentions of AI technologies. For instance, the fear that AI could become sentient and autonomously malevolent misrepresents the nature of AI, which operates within a framework of algorithms created and controlled by humans. AI systems are designed to perform specific tasks, from simple ones like filtering spam emails to more complex functions like diagnosing diseases based on medical imaging. These systems do not possess consciousness or emotional depth; they simulate understanding based on data patterns.

Moreover, the belief that AI can independently solve all problems overlooks the crucial role of human oversight in AI operations. AI is a potent tool that requires human guidance to function effectively and ethically. For example, in healthcare, AI can help predict patient risks and recommend treatments, but final decisions and ethical considerations rest securely in the hands of medical professionals. Recognizing these realities helps demystify AI's role and reframe it as a supportive technology that enhances human capabilities rather than replacing them.

Sources of Misinformation

Understanding the origins of AI myths is essential to dispelling them. Sensational media portrayals and Hollywood narratives often depict AI as a mysterious and omnipotent force, skewing public perception. While wonderfully entertaining, films and novels project futures where AI usually spirals out of control, a theme that resonates with our deepest fears of the unknown. However, these fictional accounts do not reflect actual AI development, a disciplined field of science grounded in ethics and practical applications.

The role of education in clarifying AI's realities is indispensable. Students can learn about AI's realistic capabilities, limitations, and ethical considerations by integrating AI literacy into educational curricula—from primary to higher education. For example, workshops demonstrating how AI algorithms are trained and the importance of data quality can enlighten students and adults alike, fostering a well-informed public that is both critical and appreciative of AI advancements.

Impact of Myths on Public Perception

The myths surrounding AI significantly impact its acceptance and development. Misinformation can lead to unfounded fears that hinder the integration of beneficial AI technologies in sectors that could dramatically enhance efficiency and effectiveness. For instance, the unwarranted fear of AI-driven automation might discourage businesses from adopting technologies that could streamline operations and create new opportunities for human workers. Dispelling myths and providing factual information about what AI can and cannot do is crucial to building trust and encouraging the responsible adoption of AI technologies.

Clarifying AI's Actual Capabilities

To further clarify what AI is capable of, consider using AI in customer service through chatbots and virtual assistants. These AI systems are designed to handle routine inquiries and support tickets, allowing human employees to focus on more complex customer needs. This example underscores AI's role as a supportive tool that enhances service efficiency without replacing the human touch, which is often crucial in nuanced interactions.

By understanding AI's practical applications and inherent limitations, you can more accurately appreciate its role in modern society and its potential impact on future developments. This informed perspective empowers you to leverage AI effectively in your personal and professional endeavors and engage in meaningful discussions about AI advancements' ethical and societal implications.

AI and Job Displacement: The Real Scenario

In discussions about artificial intelligence and its impact on the workforce, the narrative often swings between utopian visions and dystopian fears. On the one hand, AI is seen as a precursor of unprecedented efficiency and new job creation; on the other, it's portrayed as a force likely to usurp human jobs and exacerbate unemployment rates. However, a nuanced examination of historical trends and sector-specific impacts reveals a more complex and dynamic interplay between AI and job markets.

Historically, technological advancements have been catalysts for economic transformation, reshaping job landscapes rather than merely diminishing employment. The introduction of AI and automation in various sectors has led to job displacement in certain areas, mainly where repetitive tasks are prevalent. For instance, in manufacturing, robots can perform assembly line tasks faster and more accurately than humans. However, this displacement is often accompanied by the emergence of new job categories that require human oversight and strategic input. For example, the demand for robot maintenance technicians and AI system managers has risen in the same sectors where traditional jobs have declined. Moreover, sectors like e-commerce, which rely heavily on AI for logistics and customer service, have seen significant job creation, offsetting losses in more traditional retail settings.

The key to navigating this shift lies in reskilling and transitioning the workforce to meet evolving demands. The rapid integration of AI across industries necessitates an agile workforce equipped with skills pertinent to an AI-enhanced economy. This is where the concept of lifelong learning becomes crucial. Educational initiatives and corporate training programs are pivotal in equipping current employees with skills like data literacy, AI management, and cybersecurity, which are in high demand. Furthermore, the rise of online platforms offering AI and machine learning courses democratizes access to these essential skills, allowing individuals to adapt irrespective of their geographical or economic status.

AI emerges as a significant job creator within the tech sector and beyond. New industries and roles surround AI's diverse applications, from ethical AI development to AI-driven research. Positions like AI ethicists and data bias auditors reflect an increasing demand for roles that bridge technology with ethical governance. In healthcare, AI-assisted diagnostic tools have given rise to new support roles that facilitate the efficient use of AI in patient care management. These trends underscore AI's potential not just to replace but to create and redefine roles, contributing to a more dynamic and adaptive job market.

The transition to an AI-driven economy also underscores the critical role of policymaking and corporate responsibility. Governments and corporations must collaborate to ensure that the workforce transition is smooth and equitable. This involves creating educational programs and establishing policies that support job transitions, such as safety nets for displaced workers and incentives for businesses investing in employee retraining. Furthermore, public-private partnerships can be instrumental in developing standardized AI curricula and certifications that provide a clear pathway for individuals seeking to enter AI-driven fields.

Navigating the dual aspects of job displacement and creation in the era of AI demands a proactive approach from all stakeholders involved. By fostering an ecosystem that emphasizes continuous learning, ethical deployment, and inclusive growth, the integration of AI into workplaces can be managed in a way that not only mitigates its challenges but also maximizes its benefits. This balanced approach ensures that AI serves as a tool for economic enhancement and social progress, aligning technological advancements with human development and welfare.

AI as a Tool for Enhancement, Not Replacement

In the ongoing narrative surrounding AI, a prevalent concern is the fear of technology replacing human roles across various sectors. However, the essence and design of AI are not rooted in replacement but in augmentation—enhancing human capabilities and working alongside people to achieve more together than either could alone. When designed and implemented with a focus on augmentation, AI can significantly elevate human capacities in fields such as medicine and finance, proving to be an invaluable ally rather than a substitute.

In the medical field, AI-powered diagnostic aids exemplify how technology enhances human expertise. For instance, AI systems in radiology can analyze thousands of images rapidly, identifying patterns that might not be immediately apparent to the human eye. These systems then present their findings to radiologists, who use their judgment and experience to diagnose. This collaboration allows for higher accuracy and faster diagnosis, ultimately improving patient outcomes. Similarly, in finance, decision support systems powered by AI analyze vast quantities of market data to provide financial analysts with insights that might take days to compute manually. These insights enable analysts to make informed decisions swiftly, combining AI's computational prowess with human strategic thinking.

Collaborative AI

Developing collaborative AI systems highlights the synergy between human intuition and AI's computational power, creating increased productivity and creativity opportunities. These systems are designed to interact with human input dynamically, adapting and responding to new information in real-time. For example, AI tools can help architects generate multiple design variations based on specified criteria in architectural design. Architects can refine these designs, integrating their creative vision with AI-generated options. This partnership speeds up the design process and opens up new possibilities in innovative design that might not have been feasible solely through human efforts.

In creative industries, AI's role extends to collaborative platforms where writers, artists, and designers use AI as a co-creative partner. AI algorithms can suggest novel ideas, patterns, and layouts that professionals can refine and integrate into their projects. This collaborative process often leads to innovative outcomes that push the boundaries of traditional methods, enriching the creative landscape with new styles and perspectives that are the fruits of human-AI collaboration.

Ethical Design and Implementation

AI systems' ethical design and implementation ensure they augment rather than disrupt human roles. Ethical AI design involves creating systems that respect human dignity and choices, ensuring that AI is a supportive tool. This approach requires transparency in how AI systems operate and mechanisms for human oversight and intervention. For instance, AI systems used in hiring should be designed to eliminate biases and enhance the recruitment process, not replace human decision-making. AI can help create more diverse and competent workforces without overriding human judgment by providing decision-makers with a broader, unbiased perspective.

Ensuring that AI systems do not cause social disruption involves carefully considering how these technologies are deployed. AI should be used to automate tasks that can lead to physical strain or extreme tedium in human workers, improving workplace safety and job satisfaction. However, this automation mustn't lead to job displacement without providing avenues for affected individuals to transition into new roles or careers. Proactive measures, such as ongoing education and training programs, must be integral to AI implementation strategies in industries most affected by automation.

Case Studies

Several case studies highlight the positive aspects of AI-human collaboration. In one instance, a major automotive manufacturer implemented AI-driven robotic arms in their assembly lines. These robots were designed to work alongside human employees, handling tasks that were ergonomically challenging for humans. This collaboration increased production efficiency and reduced workplace injuries, leading to higher job satisfaction among the workforce. Employees were also trained to oversee and maintain these robotic systems, acquiring new skills that enhanced their career prospects within the industry.

Another case study involves a technology firm that developed an AI system to assist researchers by sifting through scientific papers and extracting relevant information. This AI tool supported researchers by providing quick summaries of existing research, which allowed them to focus more on experimental design and data analysis. The collaborative use of AI in this context enabled researchers to speed up the innovation cycle, contributing to faster advancements in scientific research and development.

Through these examples, it becomes evident that when appropriately designed and responsibly implemented, AI is a powerful tool for enhancing human capabilities rather than replacing them. By fostering environments where humans and AI collaborate effectively, the potential for technological and human advancement is immensely amplified, leading to a future where AI supports us in realizing our highest potential across all aspects of life.

Transparency in AI: Understanding How AI Makes Decisions

In the intricate weave of modern technology, understanding how artificial intelligence (AI) makes decisions is crucial, especially in sectors like banking and retail, where these decisions can have significant impacts. For instance, consider how AI is used in credit scoring. Traditionally, credit decisions were based on several financial indicators and human judgment. Today, AI systems analyze vast data arrays, including transaction history, shopping habits, and social media activities, to assess creditworthiness. While this can lead to more accurate assessments, it also raises questions about how these decisions are made and what factors the AI considers significant. If these AI systems are not transparent, a customer denied credit might not understand why, making it difficult to contest or improve their financial standing.

The importance of transparency in AI systems extends beyond individual understanding and touches on broader societal trust and acceptance of AI technologies. Transparency is crucial when AI applications affect public services or involve personal data. For example, in predictive policing, where AI allocates police resources based on predicted crime hotspots, a lack of transparency about how these predictions are made or what data is used can lead to community mistrust and bias concerns. Similarly, in healthcare, patients and professionals must trust that AI-driven diagnostic tools are reliable and unbiased to be widely accepted and integrated into medical practice.

Achieving transparency in AI is multifaceted and involves both technological solutions and regulatory frameworks. One technique to enhance transparency is using open-box models instead of the more opaque deep learning models. Open-box models allow users to see and understand the decision-making paths, which can be crucial in sectors where understanding AI decisions is necessary for trust and compliance. For instance, in loan approvals, an open-box AI model can provide applicants with insights into which factors influenced their credit scoring, offering an opportunity to address these factors. Another method is conducting regular algorithmic audits. These audits assess AI systems to ensure they function as intended and adhere to ethical guidelines, identifying biases or errors that could compromise decision-making.

Regulatory and ethical considerations form the backbone of AI transparency. Governments and international bodies are increasingly aware of the need to regulate AI to protect citizens and ensure fair use of technologies. For example, the European Union's General Data Protection Regulation (GDPR) includes provisions for the right to explanation, where individuals can ask for explanations of algorithmic decisions that affect them. Such regulations promote transparency and encourage developers to design AI systems with explainability in mind from the outset.

Incorporating these practices and guidelines ensures that AI systems not only support human decision-making but do so in an understandable and accountable way. This transparency is essential for building trust and fostering an environment where AI can be critically evaluated and continuously improved, ensuring its benefits are realized across all sectors of society. As AI becomes a more prominent part of everyday life, understanding and implementing transparency will be crucial for ensuring that this technology acts as a force for good, enhancing our capabilities and making informed decisions that reflect our values and ethics.

AI and Human Interaction: Complementary Forces

In the intricate dance of daily human interactions, where every gesture and word carries weight, artificial intelligence (AI) is stepping in not as a disruptor but as an enhancer of these interactions. By integrating AI tools that assist in communication and provide personalized learning environments, we are witnessing a transformation in how individuals connect and learn from every encounter. For instance, AI-powered language translation apps break down barriers that previously hindered communication between people of different linguistic backgrounds. In a classroom setting, such tools can immediately translate a teacher's instructions for non-native speakers, ensuring every student can participate fully and equally. This fosters a more inclusive environment and enhances the educational experience by allowing students to engage with learning materials in their native languages, thus deepening their understanding and retention of knowledge.

Moving beyond the classroom, AI is significantly improving customer service interactions through personalized support and predictive assistance. Consider a scenario where AI systems analyze customer behavior patterns to tailor interactions according to individual preferences and history. For example, if you frequently purchase books from an online store, AI can predict which genres you prefer and alert you when new titles are released that match your tastes. Furthermore, during live interactions, AI can guide customer service agents by providing them with customer-specific information, enabling them to resolve issues more effectively and efficiently. This personalized approach satisfies customers and empowers service agents, making the process smoother and more enjoyable for both parties.

Emotional AI

One of the most fascinating developments in AI is the emergence of emotional AI, which seeks to understand and respond to human emotions. These systems use advanced algorithms to analyze vocal inflections, facial expressions, and physiological signals to gauge an individual's emotional state. In customer service, for example, emotional AI can detect frustration or confusion in a customer's voice or facial expressions during a video call and prompt the service agent to adapt their approach, perhaps by speaking more slowly or offering additional assistance. This capability enhances the interaction and builds a deeper connection between customer and company, as customers feel genuinely understood and valued.

Future Prospects

Looking ahead, the potential applications of AI in enhancing human interactions are vast and varied. Advanced social robots equipped with emotional AI could serve as companions and caretakers for older people, engaging them in meaningful conversations and monitoring their health. These robots could provide reminders for medication, suggest activities to keep the mind and body active and offer companionship, helping to alleviate feelings of loneliness and isolation. Another promising area is the use of AI in mental health, where AI systems could detect early signs of emotional distress or depression from patterns in speech and behavior, prompting timely intervention from healthcare providers.

As we continue to explore and develop these technologies, the future of AI in human interaction looks promising and essential. AI can enhance the quality of human life by improving how we connect, understand, and care for each other. By embracing these possibilities, we can ensure that AI serves as a force for good, enhancing our interactions and relationships in ways that respect and reflect our shared humanity.

As this chapter on AI's role in enhancing human interactions concludes, we glimpse a future where AI does not replace human connection but enriches it, making every interaction more meaningful and empowered. This exploration of AI as a complementary force in our daily lives sets the stage for further discussions on how AI can be integrated responsibly and beneficially across various sectors, ensuring that technology advances hand in hand with humanity. Looking forward, the next chapter will delve into the ethical dimensions of AI, exploring how we can navigate the challenges and opportunities presented by this transformative technology, ensuring it aligns with our values and serves the greater good.

EIGHT

AI in Global Challenges and Social Good

I magine a world where the aftermath of disasters is not defined by chaos and uncertainty but by swift, organized, and efficient responses that minimize damage and accelerate recovery. In this vision, artificial intelligence (AI) is not just a tool; it is a vital ally that transforms disaster response and relief operations. The integration of AI in managing global challenges, particularly in disaster-stricken areas, offers a compelling glimpse into a future where technology and human expertise combine to foster resilience and hope amid calamity.

AI in Disaster Response: Predictions and Relief Operations

In the realm of disaster management, the predictive power of AI marks a revolutionary shift from reactive to proactive strategies. Enhanced prediction models utilize vast amounts of data from past disasters to train algorithms that can accurately predict future occurrences. This capability allows governments and humanitarian organizations to prepare more effectively, ensuring that resources are strategically allocated and readiness plans are in place before disaster strikes. For instance, AI systems can analyze weather patterns, historical disaster data, and geographical information to predict the likelihood of natural disasters such as hurricanes or earthquakes. This predictive insight enables preemptive evacuations, the prepositioning of emergency supplies, and the mobilization of rescue teams in anticipation of the event, significantly reducing the potential impact on human life and property.

The role of AI extends into the chaos that often follows disasters, where real-time data analysis becomes crucial. AI systems deployed in disaster zones collect and analyze data from multiple sources, including satellites, drones, and on-ground sensors, to provide a comprehensive view of the situation as it unfolds. This real-time analysis helps coordinate efforts by identifying the hardest-hit areas, assessing damage, and prioritizing rescue operations. For example, AI-driven drones can survey disaster-affected areas quickly, sending back vital information that helps rescue teams navigate debris and obstructions, ensuring they reach those in need swiftly and safely.

Automation in relief distribution further illustrates AI's transformative impact. AI optimizes the logistics of delivering aid, ensuring that essential supplies such as food, water, and medical kits are distributed efficiently and equitably. AI systems can devise optimal distribution plans that promptly ensure supplies reach all affected individuals by analyzing data on affected populations, infrastructure damage, and resource availability. This automated logistics support speeds up relief efforts and maximizes the effectiveness of resources—a crucial factor in the critical days following a disaster.

Moreover, AI's capacity to aid in post-disaster recovery and analysis is invaluable in building future resilience. After immediate relief efforts, AI tools assist in analyzing data collected during and after the disaster to evaluate response effectiveness and identify areas for improvement. This analysis provides insights into the strengths and weaknesses of existing disaster response strategies, informing future policies and plans. Additionally, AI can help model different recovery scenarios, assisting communities to plan their rebuilding efforts in ways that mitigate the impact of future disasters. This ongoing learning and adaptation process, driven by AI, is essential for developing more innovative, more resilient infrastructure and preparedness strategies, ultimately leading to communities better equipped to handle future challenges.

As we continue to explore AI's role in addressing global challenges, its potential to transform disaster response and relief operations is evident. By enhancing prediction accuracy, optimizing real-time operations, automating resource distribution, and facilitating post-discovery analysis, AI saves lives and empowers communities to recover and build back better. This chapter underscores the critical role of AI in crafting a world where technology is a cornerstone of global resilience and humanitarian response, paving the way for a future where no individual or community is left vulnerable in the face of disaster.

AI for Environmental Protection: Monitoring and Action

In the face of escalating environmental challenges, artificial intelligence (AI) is a pivotal tool in conserving and managing the planet's natural resources and biodiversity. AI's capacity to analyze vast and complex data sets is being harnessed to foster sustainable practices that protect the environment and ensure the well-being of future generations. AI's role in wildlife conservation, pollution control, resource conservation, and climate action demonstrates its potential to significantly support and enhance global environmental protection efforts.

In wildlife conservation, AI technologies are increasingly being deployed to monitor and protect endangered species. Automated systems equipped with AI analyze images from camera traps to track animal movements, count population numbers and monitor changes in biodiversity. This data is crucial for understanding the habits and health of species, particularly those that are threatened or endangered. AI helps predict potential threats to wildlife, such as poaching or habitat encroachment, enabling quicker interventions. Furthermore, AI-driven models can simulate various conservation strategies, providing conservationists with valuable insights into how best to protect specific species and their habitats. This proactive approach not only helps in maintaining biodiversity but also aids in the restoration of ecosystems that have been adversely affected by human activities.

Turning to the pressing pollution issue, AI is revolutionizing how environmental pollution is monitored and managed. AI systems integrate data from satellites, ground sensors, and other monitoring stations to view global pollution levels comprehensively. These systems can identify sources of pollution, such as industrial emissions or vehicular exhaust, and track their movement and concentration in real-time. This information is crucial for environmental agencies to enforce regulations and for companies to manage their emissions responsibly. Additionally, AI can predict pollution trends and model the effectiveness of different pollution control measures, guiding policymakers in crafting strategies that effectively reduce environmental contamination and protect public health.

Resource conservation is another critical area where AI is making significant inroads. Sustainable management of natural resources like water and forests ensures their availability for future generations. AI models that predict and adjust resource usage based on supply, demand, and environmental conditions are becoming increasingly common. For instance, AI-driven water management systems can optimize water usage in agriculture by analyzing weather forecasts, soil moisture levels, and crop types to irrigate fields precisely when and where water is needed. This not only conserves water but also enhances crop yields and reduces the environmental impact of farming. Similarly, AI is used in managing forestry resources, which helps map forest cover, monitor deforestation activities, and plan reforestation projects efficiently and effectively.

Lastly, the role of AI in climate action cannot be overstated. AI's ability to analyze complex climate data is crucial to understanding and combating climate change. AI models process data from climate sensors, ocean buoys, and satellites to track climate change indicators such as temperature variations, sea-level rise, and ice sheet melting. This information is vital for predicting future climate scenarios and assessing the potential impact of various climatic events. AI helps scientists and policymakers formulate more effective mitigation and adaptation strategies by providing detailed projections and simulating the outcomes of different policy decisions. For example, AI can model the impact of a proposed carbon tax on greenhouse gas emissions, helping governments understand the potential benefits and trade-offs of such policies.

In these areas, AI enhances our understanding of environmental issues and empowers us to take decisive action to preserve and restore our planet. As we continue to harness AI's capabilities, it will become a key player in our global efforts to ensure a sustainable and prosperous future for all.

Combating Global Health Crises With AI

In the face of global health crises, artificial intelligence (AI) emerges as a transformative force, redefining how we predict, manage, and respond to disease outbreaks. AI's profound impact on healthcare systems worldwide demonstrates its potential to enhance the accuracy of medical diagnoses and optimize epidemic management through improved logistics and targeted public health strategies. AI is a beacon of innovation as we navigate these challenges, driving advancements that significantly improve global health outcomes and preparedness.

AI models have become indispensable in predicting disease outbreaks by analyzing diverse data sources, including historical health records, real-time disease surveillance, and environmental factors. These models employ sophisticated algorithms to identify patterns and anomalies that may signal the emergence of a disease. For instance, by analyzing data from previous flu seasons, AI can predict future outbreaks' onset, intensity, and geographic spread, enabling health authorities to initiate preventive measures, such as vaccination campaigns, well in advance. This proactive approach helps contain the disease early and significantly reduces the burden on healthcare systems. Moreover, AI's predictive capabilities extend to identifying potential hotspots for more severe diseases like Ebola or Zika based on travel patterns and climatic conditions, ensuring that timely interventions can be targeted effectively to the most at-risk regions.

The role of AI in managing and responding to epidemics is further exemplified by its ability to optimize logistics and supply chain management of medical supplies. During an epidemic, the efficient distribution of medical resources, including vaccines, medications, and personal protective equipment, is critical. AI systems analyze real-time data on disease spread, inventory levels, and transportation networks to optimize the delivery routes and schedules for medical supplies. This ensures that resources are delivered where they are most needed and replenished promptly, maintaining an adequate supply during ongoing healthcare challenges. Additionally, AI assists in managing healthcare staffing by predicting patient inflow in hospitals and aligning staff schedules and deployment accordingly, ensuring that patient care is not compromised due to resource shortages.

Enhancing diagnostic accuracy, particularly in remote or under-served areas, is another area where AI has made significant strides. AI-driven diagnostic tools leverage machine learning algorithms to interpret medical images, such as X-rays or MRIs, with a precision that matches and, sometimes, exceeds human experts. In regions with limited access to qualified radiologists or specialists, these AI tools provide critical support in diagnosing conditions such as tuberculosis, pneumonia, or diabetic retinopathy. By integrating these tools into mobile health applications, healthcare workers in remote areas can upload medical images and receive real-time diagnostic insights, enabling prompt and accurate treatment decisions. This improves patient outcomes and democratizes access to high-quality healthcare services, bridging the gap between urban and rural healthcare provisions.

AI's integration into global health data platforms illustrates its role in synthesizing information from multiple health databases to offer a unified view of health trends and needs worldwide. These platforms collect and analyze data from public health records, hospital data systems, and other digital health sources to monitor health conditions across different populations. The insights generated by AI help identify risk factors for diseases, track health outcomes over time, and evaluate the effectiveness of public health interventions. Health authorities use this information to tailor public health policies and programs to the specific needs of their populations, enhancing the overall efficiency and effectiveness of health services. Moreover, these AI-enhanced platforms facilitate international collaboration by enabling health experts and policy-makers to share knowledge and strategies, fostering a coordinated global response to health challenges.

As AI continues to evolve, its integration into health crisis management and epidemic response systems represents a critical advancement in our ability to understand and combat diseases and foresee and forestall their impacts. This proactive and informed approach, powered by AI, is essential to safeguarding global health and ensuring communities worldwide are better prepared to face health emergencies.

AI in Human Rights: Protection and Advocacy

Artificial intelligence (AI) is emerging as a formidable ally in detecting, reporting, and remedying violations in the intricate global human rights landscape. The capacity of AI to sift through vast data sets from social media, news outlets, and eyewitness reports has revolutionized how human rights organizations oper-ate, enabling them to act swiftly and effectively. For instance, AI-driven tools scan through digital content at an unprecedented scale to identify potential human rights abuses. These tools use natural language processing to detect patterns and keywords that indicate violations, such as "unlawful detention" or "forced evic-tion," and image recognition algorithms to identify disturbing visuals from conflict zones. This rapid processing allows organiza-tions to gather evidence and respond in real time, significantly increasing the chances of preventing violence or providing timely aid to affected populations.

Moreover, AI is crucial in providing legal assistance to under-served populations. In regions where legal aid is scarce or absent,

AI-powered platforms offer guidance and resources to individuals seeking to understand their legal rights. These platforms can draft legal documents, guide individuals through legal procedures, and provide personalized legal advice based on the specific issues and jurisdiction involved. By democratizing access to legal information, AI empowers individuals to advocate for themselves and navigate complex legal systems that might otherwise be inaccessible due to financial constraints or geographic isolation.

Enhancing transparency and accountability in governance through AI is another area where significant strides are being made. AI algorithms analyze public records, financial transactions, and government actions to detect anomalies that may indicate corruption, such as irregular contract awards or unexplained asset growth among public officials. By flagging these issues, AI helps anti-corruption agencies investigate and hold those involved in unethical practices accountable. Furthermore, AI enhances the transparency of governmental operations by making data more accessible and understandable to the public, fostering a culture of accountability where citizens can more effectively oversee and influence government actions.

Supporting displaced populations is an area where AI's impact is profoundly humanitarian. AI optimizes aid distribution to refugee camps and displaced communities by predicting needs based on demographic data, past consumption rates, and ongoing monitoring of supply levels. This ensures that essential resources like food, water, and medical supplies are allocated efficiently and according to the most urgent needs. AI facilitates communication across language barriers in these diverse communities by providing real-time translation services, enabling better coordination among aid workers and smoother communication with affected individuals. Personalized AI-driven support services also help process asylum applications, schedule appointments, and provide displaced individuals with information about their rights and the services available, easing their integration into host communities and reducing the strain on overburdened asylum systems.

AI is a technological tool and a beacon of hope and empowerment through these applications, enhancing the global pursuit of justice and equity. By harnessing AI in the service of human rights, we are not only able to respond more effectively to immediate violations but also to build more transparent, just, and inclusive societies. As AI continues to evolve, its role in protecting and promoting human rights promises to expand, opening new avenues for advocacy and support that were previously unimaginable. This transformative impact underscores the potential of technology to serve humanity at its most vulnerable, reshaping how we address some of the most pressing challenges of our time.

Bridging the Digital Divide with AI Technology

In a world where digital access significantly influences educational opportunities, economic participation, and cultural engagement, artificial intelligence (AI) plays a crucial role in bridging the digital divide. This gap, which separates those with access to modern information and communication technology from those without, is a barrier to individual progression and a broader impediment to global development. By enhancing access to educational resources, improving internet connectivity, facilitating economic inclusion, and aiding in cultural preservation, AI is pivotal in ensuring that the benefits of digital transformation are universally accessible.

AI's impact in transforming educational access in underserved areas is profound. In regions with scarce educational resources, AI is more than just a supplementary tool; it becomes a critical bridge to learning opportunities. AI systems are adept at adapting educational content to local languages and literacy levels, making learning more accessible and relevant. For instance, AI-driven platforms can translate existing educational materials into multiple local dialects, breaking language barriers that often hinder education. Moreover, these platforms employ machine learning to tailor educational content to individual students' needs and learning paces. In a rural school with high student-teacher ratios, AI-powered educational tools can provide personalized tutoring, assess student progress, and adapt lessons to address learning gaps. This customized approach ensures that students in remote areas receive a quality education attuned to their unique learning contexts and challenges.

Improving internet connectivity in remote areas is another area where AI is making significant strides. AI utilizes predictive models to analyze data points such as geographic terrain, population density, and existing network infrastructure to identify the most effective locations for new internet infrastructure. This optimization ensures that investments in connectivity yield the highest possible returns regarding expanded access. For communities in remote or rural areas, enhanced connectivity means better access to online resources, telemedicine, e-commerce, and digital communication tools, which are essential for participating in the modern digital economy. Moreover, AI-driven network management tools continuously monitor and adjust the network performance, ensuring that connectivity remains robust and interruptions are minimized, thereby enhancing the reliability of digital services.

AI-driven economic inclusion is transforming how financial services are delivered to underserved communities. In regions lacking traditional banking infrastructure, AI enables fintech solutions that offer accessible financial services via mobile platforms. By analyzing transaction data, user behavior, and economic trends,

AI can tailor financial products such as microloans, insurance, and savings programs to the specific needs of these communities. For small-scale entrepreneurs in these areas, access to credit via AI-driven platforms can be a game-changer, providing the capital needed to start and grow businesses. Additionally, AI helps assess credit risk more accurately, reduce default rates, and make financial systems more sustainable and resilient. This tailored and inclusive approach boosts local economies and promotes economic stability and growth.

Cultural preservation and digitization are crucial in maintaining the rich tapestry of global heritage, and AI is at the forefront of these efforts. Through digital archiving, AI assists in preserving cultural artifacts, texts, and traditions, converting them into digital formats accessible to a global audience. This process safeguards cultural heritage against the ravages of time and environmental threats and makes it available for educational purposes and cultural exchange. AI enhances these digital archives by offering virtual tours, 3D reconstructions, and interactive experiences that bring cultural heritage to life. For communities whose histories are orally transmitted or at risk of being forgotten, AI-driven preservation projects ensure that their stories and traditions are recorded and shared with the world, fostering a greater appreciation and understanding of global cultural diversity.

As AI continues to evolve, its role in bridging the digital divide is becoming increasingly evident. By making digital tools and resources more accessible, AI enhances individual opportunities and contributes to the collective progress of global society. This chapter highlights the transformative impact of AI in making digital equity a reachable goal, paving the way for a future where technology serves as a universal enabler of human potential.

Transition to the Next Chapter

The exploration of AI's role in bridging the digital divide sets the stage for a deeper examination of how AI shapes the future of governance and public policy. In the next chapter, we delve into AI's impact on enhancing governmental transparency, efficiency, and citizen engagement, revealing how technology transforms individual lives and redefines public administration and governance structures.

NINE

The Future of AI—Trends and Innovations

Imagine stepping into a realm where the boundaries of computational power are not just pushed but wholly redefined. This is the landscape of quantum computing, intertwined with the advanced capabilities of artificial intelligence (AI), heralding a new era of technological synergy. As we explore this junction of profound computational evolution, we unlock potentials once confined to science fiction pages. Quantum computing, with its ability to handle complex problems at unprecedented speeds, is not just enhancing AI; it's revolutionizing the fabric of problem-solving methodologies and security paradigms in the digital age.

Quantum Computing and AI: The Next Frontier

The fusion of quantum computing and AI marks a pivotal shift in our data analysis and problem-solving approach. Traditional computers power current AI systems and process information in binary bits—either zeros or ones. Quantum computers, however, utilize quantum bits or qubits, which can represent both zero and one simultaneously, thanks to the phenomenon known as super-position. This capability allows quantum computers to process vast amounts of data at speeds unattainable by classical computers. For AI, this enhanced computational power means algorithms can analyze data, learn patterns, and make decisions much more quickly and efficiently. This is not just an incremental improve-ment; it's a monumental leap that could shorten the time needed to solve complex problems—from optimizing large-scale logistics to discovering new pharmaceuticals—from years to minutes.

The development of quantum machine learning algorithms is revolutionizing how we approach AI's learning capabilities. These algorithms can dramatically enhance pattern recognition and data analysis processes, which are core aspects of machine learning. For instance, quantum algorithms are particularly adept at handling unstructured data, which makes up the vast majority of data in the world, from images and videos to human language. The ability to quickly sift through and make sense of this data could lead to breakthroughs in everything from natural language processing to real-time translation systems that more accurately understand and replicate human nuances.

The impact of quantum computing extends deeply into cryptog-raphy and security. Traditional encryption methods, which secure everything from online transactions to personal communications, rely on complex mathematical problems that are time-consuming for classical computers to solve. However, quantum computers could solve these problems more swiftly, potentially rendering current encryption methods obsolete. This challenge necessitates the development of quantum-resistant encryption techniques—new security protocols that can withstand the power of quantum computing. This ongoing cat-and-mouse game between encryp-tion and decryption highlights the dynamic nature of technological advancement and the continuous need for innovation in cybersecurity.

Quantum computing also presents significant technical and ethical challenges despite its immense potential. The hardware required to build a stable quantum computer is extraordinarily complex and sensitive, requiring conditions that are difficult to maintain, such as extremely low temperatures. Moreover, the ethical implications of such powerful computing technology—from privacy concerns to the potential for creating new technological access divides—are profound. As we stand on the brink of this quantum leap, the need for robust ethical frameworks and global cooperation to manage these advances becomes increasingly apparent. Ensuring that the benefits of quantum AI are accessible and beneficial to all requires thoughtful governance and a commitment to equity.

As we contemplate the horizon of AI's future, marked by the integration of quantum computing, it becomes clear that we are not just observers but active participants in this next great chapter of technological evolution. The convergence of these powerful technologies promises to redefine what is possible, pushing the boundaries of science, business, and society. The journey ahead is as exciting as it is uncertain, and staying informed and engaged is crucial as we navigate this uncharted territory.

AI in Space Exploration: Unmanned Missions

The realm of outer space, once the canvas for celestial mysteries and cosmic marvels, is now increasingly accessible thanks to the strides made in artificial intelligence (AI). As we push the boundaries of human knowledge and venture further into the unknown, AI plays a pivotal role in shaping the future of space exploration. Autonomous spacecraft, for instance, represents a significant leap in how we approach missions beyond Earth. These sophisticated vessels, equipped with AI systems, are designed to operate with minimal human intervention, making decisions and navigating the vast expanse of space on their own. This autonomy is crucial, particularly for missions to distant planets or asteroids, where the delay in communication with Earth can span several minutes to hours. AI enables these spacecraft to perform critical tasks, from maneuvering to avoid collisions with space debris to adjusting their trajectory based on real-time astronomical data. The ability of these spacecraft to self-manage their operations not only enhances the efficiency of space missions but also reduces the risks to human life by delegating hazardous tasks to unmanned probes.

Deep space communication is another area where AI is making a profound impact. The vast distances of space can cause significant delays and disruptions in signals, posing a challenge to the control and coordination of space missions. AI algorithms optimize the processing and transmission of data across these immense distances, ensuring that communication between spacecraft and Earth remains as seamless as possible. These AI systems can adjust transmission power and frequencies in response to fluctuating conditions in space, such as solar flares or electromagnetic storms, which can interfere with signal clarity. Moreover, AI enhances the analysis of incoming data from space missions, swiftly decoding and prioritizing critical information for immediate decisions. This capability streamlines operations and ensures that scientists and engineers on Earth can quickly respond to any anomalies or opportunities that the spacecraft encounters.

AI-powered rovers and probes have particularly revolutionized the exploration of other planets. These machines are equipped with various sensors and instruments that collect data on geological and atmospheric conditions, but AI turns this data into valuable insights. On planets like Mars, AI-driven rovers analyze soil samples and atmospheric data to search for signs of water, other resources, and, potentially, signs of past or present life. The AI systems onboard these rovers are trained to recognize geological patterns and anomalies that might indicate the presence of essential resources or interesting scientific phenomena. This autonomous capability allows rovers to explore hostile environments more effectively and make discoveries faster than human-operated vehicles. The data these rovers collect helps scientists on Earth understand the conditions of other planets, paving the way for future manned missions and long-term colonization plans.

AI is also transforming the management and coordination of satellite swarms. These satellites, often small and numbering in the dozens or even hundreds, work in concert to perform complex tasks ranging from detailed Earth observation to tracking natural disasters and mapping global weather patterns. AI algorithms orchestrate the movements and functions of these satellite swarms, ensuring they operate in harmony and avoid collisions. This coordination is crucial, especially as the number of satellites in orbit grows, raising the potential for space traffic issues. AI-driven automation allows these satellites to adjust their orbits in response to mission requirements or to move out of the way of incoming space debris. Additionally, the collective data gathered by these swarms offers unprecedented insights into environmental changes, urban development, and other global phenomena, showcasing AI's power to enhance our understanding of Earth and outer space.

The Evolution of AI in Autonomous Vehicles

As technology propels forward, the realm of autonomous vehicles (AVs) exemplifies a cutting-edge fusion of AI capabilities, where cars navigate and make decisions without human input. The march toward Level 5 autonomy signifies a future where vehicles operate independently under all driving conditions. This pinnacle of autonomous technology means cars will not only handle all aspects of driving. Still, it will do so in ways that adapt to varying environments and conditions without human intervention. Imagine a scenario where your vehicle can take you from a sunny suburban street through a foggy mountain pass and into a chaotic urban setting, all while you read a book or catch up on emails. Achieving this level of sophistication involves intricate layers of machine-learning models that process data from sensors and cameras to understand and interact with the world in real time.

The pathway to full autonomy is fraught with complex challenges, not least the ethical decision-making required during unavoidable accidents. Here, AI must navigate moral dilemmas that would challenge even the most seasoned human drivers. Consider the trolley problem, a classic ethical dilemma that questions whether one should act to change the course of a runaway trolley if it means harming one person to save several others. In autonomous vehicles, AI may face similar split-second decisions, such as avoiding a sudden obstacle by swerving into a barrier or staying the course to protect occupants but endangering a pedestrian. Programming AI to handle these decisions involves technical capabilities and an underlying ethical framework that guides its choices, often based on extensive simulations and ethical guide-lines developed in collaboration with policymakers, ethicists, and the public.

Integration with intelligent city infrastructure represents another leap forward for autonomous vehicles. In smart cities, everything from traffic lights to road sensors is interconnected, forming a responsive network that communicates with vehicles to optimize traffic flow and enhance safety. AI in autonomous cars uses this wealth of data to make informed decisions about routes and speeds, significantly reducing congestion and smoothing out traffic patterns. For instance, if an intelligent traffic system identi-fies a congestion buildup at one intersection, it can reroute vehicles in real time, easing traffic pressure and reducing the likelihood of accidents. This harmonious integration makes transportation more efficient and reduces the environmental impact of idling engines and stop-start driving, commonly seen in heavy traffic.

Safety and regulation are paramount to the widespread adoption of autonomous vehicles. As these vehicles approach the roads, ensuring their reliability and safety for all road users becomes critical. Regulatory bodies worldwide are crafting frameworks to govern the deployment of AVs, focusing on standards and practices that safeguard public safety. Continuous improvements in AI algorithms enhance the vehicle's ability to detect and react to unexpected conditions, from sudden weather changes to erratic human-driven cars. The ongoing evolution of these regulations seeks to keep pace with technological advancements, ensuring that safety protocols are robust and comprehensive. Additionally, testing protocols are becoming more rigorous, involving millions of simulated driving miles to expose the AI to a wide range of scenarios, ensuring the vehicle can handle real-world complexities.

As AI continues to evolve within the autonomous vehicle sector, the implications for our daily lives and urban landscapes are profound. These advancements promise enhanced convenience and safety and a reimagining of transportation norms that have stood for decades. With each incremental step toward full autonomy, the integration of AI in our vehicles and cities paints a bold vision of the future—one where our journeys are not just about reaching destinations but doing so in a way that is fundamentally aligned with broader goals of efficiency, safety, and sustainability.

Next-Generation Robotics and AI

In the evolving robotics landscape, the emergence of collaborative robots, commonly known as robots, marks a significant step toward harmonious human-robot interactions. Robots are designed to work alongside humans in various settings, from intricate assembly lines in manufacturing plants to the comfort of domestic environments. Unlike traditional robots that operate in isolation, robots are equipped with advanced sensors and AI systems that allow them to understand and adapt to human presence and activities. This capability transforms workplace dynamics by enabling robots and humans to share tasks, combining human dexterity and decision-making with robotic precision and endurance.

In industrial settings, robots enhance safety and efficiency by taking on repetitive or physically demanding tasks, reducing the risk of injuries, and allowing human workers to focus on more complex aspects of production. For example, in automobile manufacturing, robots assist with installing heavy components and meticulously handling and positioning parts, while human technicians oversee the process and make adjustments as necessary. This collaboration speeds up the assembly process and minimizes errors and material waste, leading to higher productivity and improved worker satisfaction. In homes, robots extend their utility by performing daily chores, learning household preferences, and adjusting their functions to support a wide range of tasks, from cleaning to assisting with mobility for elderly or disabled individuals.

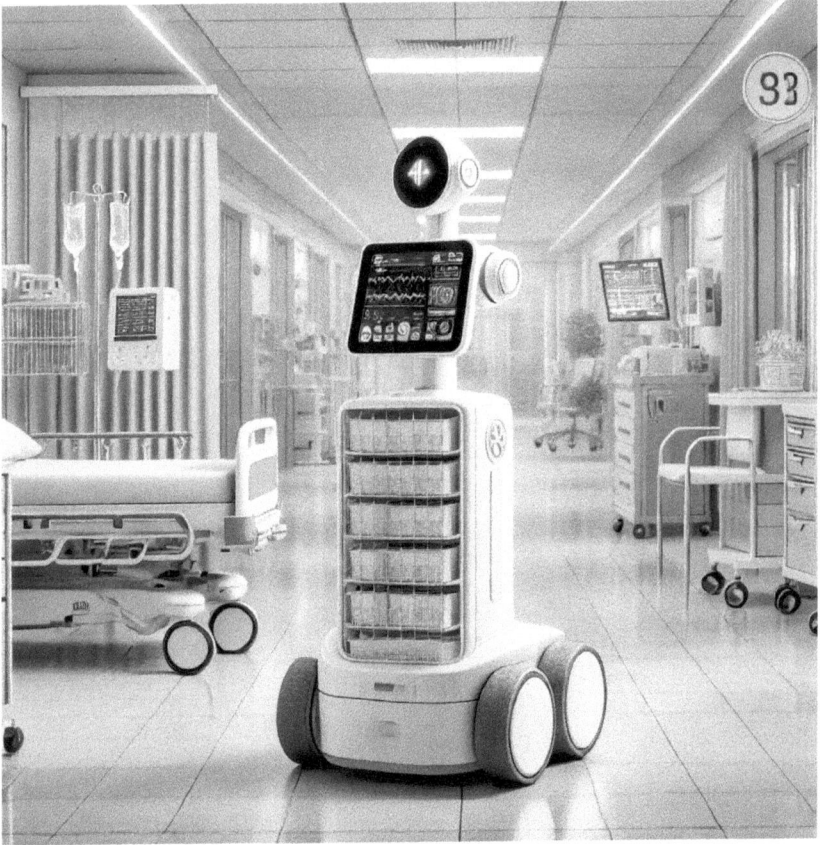

The realm of robotics is also witnessing significant advancements in AI-driven manipulation, enabling robots to perform tasks that require exceptional precision and delicacy. In the medical field, robotic systems are increasingly used in surgeries, where they can manipulate instruments with a precision that surpasses human capabilities. These robotic systems, guided by AI, can perform complex procedures such as delicate eye surgeries or minimally invasive operations, where precision is critical for patient safety and recovery. The AI algorithms that control these robots are trained on vast datasets of surgical procedures, allowing them to learn and replicate the most effective techniques. Beyond healthcare, robots equipped with advanced manipulation capabilities are transforming industries that handle fragile goods, such as electronics or glass manufacturing, where precision handling can significantly reduce the risk of damage and improve the quality of the final products.

Emotional intelligence in robots is another frontier in AI research, enhancing how robots understand and respond to human emotions. This facet of AI allows robots to interpret verbal cues, facial expressions, and body language to gauge human emotions and react appropriately. In customer service, emotionally intelligent robots can assess customers' moods and tailor their interactions, providing empathetic responses and adjusting their communication style to improve the service experience. In caregiving, robots with emotional AI can offer companionship and support to individuals with mental health issues or cognitive impairments, responding to emotional cues to provide comfort or alert human caregivers when necessary. Integrating emotional intelligence into robots opens new avenues for AI applications in fields where emotional responsiveness is critical, enhancing the quality of interaction and support provided by robotic systems.

The development of self-learning robots represents a paradigm shift in robotics, allowing machines to adapt and improve their performance autonomously. Unlike traditional robots, which require explicit programming for each task, next-generation robots utilize machine learning algorithms to learn from experience and refine their operations over time. This capability enables robots to tackle a broader range of activities and adjust to new environments without human intervention. For example, a self-learning robot in a logistics warehouse can improve its path optimization strategies by analyzing data from its daily operations, identifying patterns and obstacles, and adjusting its routes to minimize travel time and energy consumption. Similarly, robots in dynamic environments, such as disaster sites or exploratory missions, can adapt their strategies based on real-time data, enhancing their effectiveness in unpredictable conditions. This ability to learn and adjust extends the functionality of robots and reduces the need for frequent reprogramming, making robotic systems more versatile and cost-effective.

As we delve deeper into next-generation robotics and AI capabilities, these technologies' potential applications and benefits continue to expand, reshaping industries and daily life. The collaboration between humans and robots, guided by sophisticated AI, enhances productivity and safety and redefines the possibilities that robots can achieve. As these technologies evolve, they promise to unlock new potentials in automation, precision, and human-robot interaction, heralding a future where robots are integral and beneficial companions in our professional and personal lives.

Ethical AI Futures: Governing the Unknown

As artificial intelligence (AI) continues to evolve, permeating every sector of society, the need for proactive regulation has never been more critical. As a member of this technologically advanced society, you may wonder how these technologies are kept in check. Proactive regulatory frameworks are crucial because they prepare the groundwork for rapid AI advancements while ensuring these innovations align with ethical considerations. Such regulations are not merely reactive measures; they anticipate future developments and challenges, ensuring that AI progresses in ways that benefit society without compromising moral values or causing unintended harm. For instance, consider AI in healthcare, where regulatory frameworks could specify standards for patient data usage, consent processes, and the transparency of AI-driven diagnostic decisions. These proactive measures ensure that, as AI technologies evolve, they adhere to standards that protect patient rights and promote trust in AI applications.

Transparency in AI development is another cornerstone of ethical AI futures. As AI systems become more integral to our lives, understanding how these systems make decisions is paramount. Greater transparency helps demystify AI processes for the public, fostering trust and acceptance. For example, if an AI system denies your loan application, understanding the factors influencing this decision can provide insights and a pathway to rectify any issues. This level of transparency is achieved through clear documentation of AI processes, open communication about the data used, and the logic behind AI decisions. Efforts to enhance transparency help build trust and empower users to interact with AI with greater confidence and understanding.

The push for global AI governance is a response to the universal impact of AI technologies, transcending national boundaries and cultural contexts. Establishing global norms and policies for AI that respect diverse cultural and ethical values while promoting safe and beneficial AI use is essential. This global governance framework ensures that AI development is harmonized worldwide, preventing disparities that could lead to misuse or uneven benefits from AI technologies. For example, international agreements on AI use in military operations or global standards for autonomous vehicle safety can help ensure these technologies are used responsibly worldwide. Such efforts require collaboration among nations, industries, and communities to create a comprehensive governance structure that addresses the varied implications of AI, from privacy and security to fairness and inclusiveness.

Balancing innovation with accountability in AI development presents unique challenges. While fostering innovation is crucial for technological progress, ensuring that such innovations are accountable for their impacts on society is equally essential. Strategies to maintain this balance include implementing robust testing phases for AI systems, where they are evaluated for functionality, ethical implications, and potential social impacts. For instance, before deploying an AI system in public spaces, it could undergo simulations to identify any biases in its behavior or potential privacy infringements. Moreover, creating mechanisms for accountability, such as audit trails and impact assessments, ensures that AI developers and users remain responsible for the outcomes of AI systems. These strategies help cultivate a culture of responsibility and caution around AI innovations, ensuring that technological advances do not outpace our ability to manage them responsibly.

As AI becomes an ever-more critical component of modern existence, shaping everything from our devices to global infrastructure, the discussions and decisions around its ethical governance will shape the legacy of our era. By fostering an environment where innovation is balanced with conscientious regulation, transparency, and global cooperation, we ensure that AI serves as a force for good, propelling humanity toward a future where technology amplifies our potential without compromising our values. As we close this chapter on the ethical futures of AI, we look toward a horizon rich with opportunity, guided by the principles of equitable and responsible AI use. The next chapter will explore practical applications of AI, bringing these high-level concepts into concrete examples of how AI is being integrated into various sectors of society and demonstrating the tangible benefits of this remarkable technology.

TEN

Building a Career in AI

Imagine stepping into a future where your skillset not only opens doors to revolutionary fields but also positions you at the forefront of technological evolution. This is the promising horizon for those venturing into a career in artificial intelligence (AI). As AI continues to weave its intricate tapestry across various sectors, understanding the mosaic of skills required to thrive in this dynamic field becomes imperative. This chapter dedicates itself to unfolding the essential skills and continuous learning pathways that can propel you into the exhilarating AI world, whether you're a budding enthusiast or a seasoned professional.

Skills Needed for a Career in AI

Embarking on a career in AI is akin to preparing for a marathon; it requires a robust foundation in core technical skills, a keen understanding of statistical and mathematical principles, proficiency in specialized software and tools, and an unwavering commitment to continuous learning. As you navigate AI's complex and ever-evolving landscape, these elements are accessories and essential gears in your toolkit.

Core Technical Skills

The backbone of any AI professional's expertise lies in their technical prowess. Programming languages like Python and R have emerged as the linchpins in developing AI applications. Its simplicity and readability, coupled with an extensive range of libraries such as NumPy for numerical computation or TensorFlow for machine learning, offers a versatile base for building and experimenting with AI models. Meanwhile, R provides a robust environment for statistical analysis and data visualization, making it invaluable for data-driven insights in AI projects. Beyond languages, a deep understanding of machine learning algorithms—from supervised and unsupervised to neural networks and reinforcement learning—is crucial. These algorithms are the engines of AI, driving everything from natural language processing tools that power voice assistants to recommendation systems that curate your digital experiences.

Statistical and Mathematical Acumen

At its core, AI is deeply rooted in statistics and mathematics. These disciplines form the bedrock on which AI models are built and optimized. Statistical knowledge enables you to understand data distributions, variance, and probability—essential for training accurate models. Meanwhile, a firm grasp of linear algebra, calculus, and optimization is indispensable for developing algorithms to learn and make predictions. For instance, linear algebra underpins the operations of neural networks, while calculus helps optimize these networks during the training process through backpropagation. Whether you are tweaking an algorithm to improve its accuracy or interpreting the output of a deep learning model, your mathematical and statistical foundation will be vital in making informed decisions.

Software and Tool Proficiency

In the toolbox of an AI professional, software proficiency is non-negotiable. Platforms like TensorFlow and Karas facilitate the design, testing, and deployment of machine learning models with efficiency and scalability. TensorFlow, developed by Google, offers a comprehensive ecosystem that supports research and production, providing a library of prebuilt functions and models that can be adapted to new projects. Karas, often used with TensorFlow, simplifies the coding required for neural networks, making the technology more accessible to beginners without sacrificing the power needed for complex projects. Mastery of these tools empowers you to bring AI applications from concept to reality, whether in academic research or commercial development.

Continuous Learning

AI is characterized by rapid evolution and innovation, making continuous learning beneficial and necessary for those who wish to remain relevant. The landscape of AI today may be radically different in just a few years, with new algorithms, technologies, and methodologies emerging at a breakneck pace. Engaging with ongoing education through online courses, workshops, and certifications keeps your knowledge base fresh and competitive. Platforms like Coursera and Udemy offer classes on everything from basic AI principles to cutting-edge techniques in deep learning and AI ethics. Additionally, attending conferences and participating in workshops helps us learn from leading experts, network with peers, share knowledge, and collaborate on projects.

This continuous learning journey ensures that your skills remain sharp and your career trajectory stays upward, aligned with the advancements in the field.

As you weave through the complexities of AI, armed with the right skills and a commitment to ongoing education, you stand ready to participate in and actively shape this transformative field's future. The opportunities in AI are as vast as the technology itself, and with the proper preparation, you can carve out a successful and fulfilling career in this exciting domain.

Finding Your Niche in the AI Industry

Navigating the vast expanse of the AI industry can feel akin to exploring a dense, uncharted forest—each path leads to a different sector with unique challenges and rewards. AI's application spans various industries, from enhancing diagnostic accuracy in health-care to revolutionizing customer interactions in finance, driving automation in automotive manufacturing, and crafting personalized entertainment experiences. This diversity highlights AI's ubiquity and offers many opportunities to find a niche that resonates with your passions and skills.

In healthcare, AI's impact is profound, with algorithms that can analyze medical imaging faster and often more accurately than human eyes, helping diagnose diseases such as cancer early and improving patient outcomes. Suppose your passion lies in making a direct impact on people's lives. In that case, this sector allows you to work on cutting-edge technology and contribute to lifesaving innovations. Conversely, the finance sector utilizes AI to enhance everything from fraud detection systems to algorithmic trading and personalized banking services. Here, the thrill lies in the fast-paced environment and the significant impact your work could have on the economic landscape.

Meanwhile, the automotive industry is at the forefront of integrating AI with the development of autonomous vehicles. If you're drawn to engineering and robotics, this field offers a dynamic environment where mechanical engineering meets innovative AI solutions. Lastly, the entertainment industry uses AI to customize viewing recommendations, create music, or even script movies, representing a fusion of creativity and technology—an ideal arena for those who wish to transform traditional media landscapes.

Understanding the different roles within these sectors is crucial to identifying where you might fit. For instance, data scientists in AI extract insights from complex data sets and drive decision-making processes with predictive analytics. This role requires a keen analytical mind and a strong foundation in statistics. On the other hand, AI engineers focus on building and maintaining scalable AI models, often requiring robust programming skills and a deep understanding of AI frameworks. Becoming an AI researcher might suit you if you are inclined toward AI's academic or theoretical aspects. Researchers push the boundaries of what AI can achieve, working on innovations like quantum computing integration or exploring new paradigms in machine learning. Alternatively, business intelligence developers use AI to transform data into actionable business insights, making strategic decisions that drive business growth. Each role plays a pivotal part in the AI ecosystem, and understanding these can help you align your career trajectory with your interests and expertise.

Another critical aspect is the working environment, which varies significantly between startups and large corporations. Startups offer a dynamic atmosphere where wearing multiple hats is often necessary, providing you with a breadth of experience and a steep learning curve. The pace is fast, and the impact of your work is immediately visible, which can be immensely satisfying. However, this can also mean a less structured environment with limited resources. Large corporations, by contrast, offer more specialized roles, often with access to cutting-edge technologies and more significant resources. The trade-off here includes navigating complex hierarchies and potentially slower decision-making processes. Your choice between these environments should align with your work style and career goals, whether you thrive in a fluid, fast-paced atmosphere or prefer a structured, steady approach.

Lastly, the ethical implications of AI development cannot be over-stated. As you consider your niche, reflect on the moral dimensions of your work. AI's potential to influence society is immense, and with this power comes the responsibility to ensure that the technology is developed and implemented responsibly. Whether it's advocating for fairness in AI algorithms to prevent biases in automated decision-making or ensuring privacy and security in AI applications, your role in shaping ethical AI practices is crucial. Aligning your career with these values contributes to the respon-sible growth of technology and ensures that your work positively impacts society.

As you explore these various paths, remember that your niche in AI is not just about adapting to an existing role but about carving out a space where your passions and skills meet the industry's needs. This alignment is crucial for a fulfilling and impactful career in AI, where the opportunities to innovate and make a difference are boundless.

Networking and Professional Development in AI

In the vibrant and ever-evolving field of artificial intelligence, engaging with professional communities is not just beneficial—it's crucial for fostering growth, innovation, and collaboration. Whether you attend global conferences, participate in specialized

AI seminars, or join online forums, these platforms offer invaluable opportunities for exchanging ideas and staying abreast of the latest developments in AI. Imagine immersing yourself in an environment where every conversation and workshop opens up new avenues of knowledge and collaboration. Such interactions can spark new ideas, lead to collaborations, or even pivot your career direction by exposing you to different perspectives and cutting-edge research.

Furthermore, the role of mentorship in AI must be balanced. Navigating a career in such a dynamic field can be daunting, and having a mentor to guide you through the complexities of AI can be immensely beneficial. Mentors provide technical guidance and career advice, helping you navigate challenges and seize opportunities. They can introduce you to industry networks, recommend learning resources, and provide insights based on their experiences. Engaging with a mentor can be as simple as reaching out to a professional you admire at a conference, participating in formal mentorship programs hosted by professional AI organizations, or connecting with thought leaders on professional social media platforms. The relationships you cultivate with mentors can profoundly impact your professional development and career trajectory in AI.

Building a professional brand in the AI industry is another strategic step in carving out your niche and establishing yourself as a thought leader. Start by maintaining an active LinkedIn profile where you share your achievements, projects, and insights into AI. Contributing to open-source AI projects hones your skills and showcases your commitment and practical expertise to potential employers or collaborators. Moreover, blogging about AI topics or speaking at industry events can elevate your profile. These activities help you articulate your knowledge and viewpoints, positioning you as a knowledgeable person in the field.

They also provide discussion content and can attract opportunities for collaborations and job offers.

Leveraging social media effectively is integral to staying connected with the AI community and keeping up-to-date with industry trends. Platforms like Twitter and LinkedIn are not just for networking; they are rich resources for learning. Follow leading AI researchers, industry pioneers, and professional groups to gain insights into their work and the latest AI advancements. Engage actively by participating in discussions, sharing relevant content, and showcasing your projects. This visibility can be instrumental in establishing your professional identity in the AI community.

By actively engaging in these professional development activities, you will enhance your knowledge and skills and establish a robust network to propel your AI career. These connections can lead to job opportunities, collaborative projects, and continuous learning —all essential for a successful career in this dynamic field.

Navigating through the complexities of professional development in AI, from networking at conferences to engaging with mentors and building a solid professional brand, sets the stage for a thriving career. This exploration enhances your understanding of AI and aligns your professional pursuits with your aspirations and values. Integrating these strategies into your career development will pave the way for individual success and meaningful contributions to AI. Next, we'll delve into how these foundational elements of networking and development play out in real-world scenarios, connecting theory with practice in the vibrant world of AI.

Spread the Word: AI Can Be the Key to Human Progress

Technology can profoundly affect human beings, empowering them to be more mindful, sustainable, and complete. You have seen how AI can boost your enjoyment of hobbies, foster deeper learning in various subjects, and keep you and your loved ones safer in your home.

I hope you have seen that many of the most common fears about AI are unfounded… and that we can coexist alongside this technology, using its unique abilities to transition and keep up with evolving demands. If you feel more confident about how you can make AI work for you instead of against you and its potential for fostering social good inspires you, please share your thoughts with others.

LEAVE A REVIEW!

It will only take a sentence to show others how to embrace a future in which AI helps them achieve tremendous success, purpose, and quality of life. That's one thing that is still exclusively a human sensation—enjoyment!

Conclusion

As we stand on the brink of a revolution, it's crucial to recognize the transformative potential of artificial intelligence, which we have explored throughout this journey. AI is not merely a technological advancement but a catalyst for enhancement across all facets of life. AI's role is pivotal, from boosting productivity and fostering creativity in the workplace to advancing the frontiers of healthcare and education. It addresses monumental global challenges through more thoughtful disaster response, proactive environmental conservation, or bridging the formidable digital divide. This book aims to unveil AI's vast capabilities and illustrate its profound impact on our daily lives and global society.

However, with great power comes great responsibility. The development of ethical AI is not an option but a necessity. This book has underscored the importance of creating AI systems prioritizing privacy, transparency, and inclusivity. Maintaining a vigilant and ongoing dialogue about these ethical considerations is imperative as we continue integrating AI into society's fabric.

Only through thoughtful regulation and mindful innovation can we ensure that AI evolves as a force that benefits all of humanity.

It's essential to dispel the notion that AI is a field reserved for technologists and scientists alone. AI is ubiquitous and provides opportunities for teachers, artists, entrepreneurs, and beyond. Whether you're a job seeker looking to upskill, a business owner curious about AI for competitive advantage, or a parent interested in AI's educational tools, there's a place for you in this evolving landscape. This book has strived to demystify AI, making it accessible and relatable, ensuring that anyone interested can understand and engage with this technology.

I encourage you, the reader, to not just be a bystander but to actively engage with AI. Explore educational resources, consider how AI might be leveraged in your profession, or initiate personal projects. Your unique perspective and skills can harness many opportunities for innovation and problem-solving through AI.

AI's horizon is expansive, with advancements like quantum computing and autonomous vehicles poised to redefine what's possible. Preparing for these changes is crucial—we must educate ourselves, participate in policymaking, and steer the ethical discourse to shape a future where AI and humans coexist beneficially.

Let us embrace AI as a force for good. A tool that, when used responsibly, enhances our capabilities, improves our quality of life, and addresses some of the most pressing challenges of our time. I urge you to participate in shaping this future. Engage in ethical discussions, advocate for inclusive and fair policies, and contribute to a community that values diverse perspectives in guiding AI development.

For those keen on delving more deeply, I encourage you to join AI communities, contribute to open-source projects, and share your insights and experiences. Your involvement can significantly affect how AI evolves and impacts our world.

Thank you for joining me on this enlightening journey through the world of AI. For further exploration, I recommend various resources, from online courses offered by platforms like Coursera and Udemy to insightful podcasts and publications that delve into the nuances of AI advancements.

In conclusion, let's look forward with optimism. With responsible development and thoughtful application, AI promises to revolutionize our current landscape and offer us a future where technology and humanity advance hand in hand toward a more efficient, equitable, and sustainable world. Let's seize AI's immense opportunities and ensure this technological revolution benefits everyone everywhere.

References

Twenty-four best AI tools for special education teachers in 2024. (2024, March 1). MTI: Professional Development Courses & Graduate Ce for Teachers. https://www. midwestteachersinstitute.org/special-education-ai-tools/

Administrator. (2021, December 7). Forty-three *examples of artificial intelligence in education.* University of San Diego Online Degrees. https://onlinedegrees. sandiego.edu/artificial-intelligence-education/

Artificial Intelligence and Conservation. (n.d.). World Wildlife Fund. https://www. worldwildlife.org/pages/artificial-intelligence-and-conservation

Artificial intelligence in education: Teachers' opinions on AI in the classroom – Forbes advisor. (n.d.). https://www.forbes.com/advisor/education/it-and-tech/artifi cial-intelligence-in-school/

Blackman, R. (2020, October 15). A practical guide to building ethical AI. *Harvard Business Review.* https://hbr.org/2020/10/a-practical-guide-to-building-ethical-ai

Clark, E. (n.d.). *Personal branding with AI and marketing automation.* Forbes. https:// www.forbes.com/sites/elijahclark/2023/12/04/personal-branding-with-ai-and-marketing-automation/

David Nolan, Hajira Maryam & Michael Kleinman. (2024, January 16). *The urgent but difficult task of regulating artificial intelligence.* Amnesty International. https:// www.amnesty.org/en/latest/campaigns/2024/01/the-urgent-but-difficult-task-of-regulating-artificial-intelligence/

Dice Staff. (2024, April 18). *Ai jobs demand analysis: Current trends and future outlook in artificial intelligence careers.* Dice Insights. https://www.dice.com/career-advice/ai-jobs-demand-analysis-current-trends-and-future-outlook

Ehrlich, C. (2023, August 14). *Ai in action: Improving customer experiences.* CMSWire. https://www.cmswire.com/customer-experience/ai-in-customer-experience-5-companies-tangible-results/

Eric. (n.d.). *Rubric for evaluating AI tools for schools.* https://www.controlaltachieve. com/2024/04/rubric-for-evaluating-ai-tools-for.html

Global AI law and policy tracker. (2024, February). *Iapp.* https://iapp.org/ resources/article/global-ai-legislation-tracker/

Goel, V. (2023, October 27). *Unmasking AI myths in business: Navigating the realities of artificial intelligence.* eLearning Industry. https://elearningindustry.com/unmask ing-ai-myths-in-business-navigating-the-realities-of-artificial-intelligence

Hare, N. (2023, November 17). *How small businesses use AI—And how your business can benefit too.* Forbes. https://www.forbes.com/sites/allbusiness/2023/11/17/how-small-businesses-are-using-ai-and-how-your-business-can-benefit-too/

Henkin, D. (n.d.). *Orchestrating the future—AI in the music industry.* Forbes. https://www.forbes.com/sites/davidhenkin/2023/12/05/orchestrating-the-future-ai-in-the-music-industry/

https://www.artshub.com.au/news/opinions-analysis/exploring-the-ethics-of-artificial-intelligence-in-art-2694121/. (2024, January 15). *Exploring the ethics of Artificial Intelligence in art.* https://www.artshub.com.au/news/opinions-analysis/exploring-the-ethics-of-artificial-intelligence-in-art-2694121/

Know more about the rise of AI meditation apps and their advantages. (n.d.). KnowledgeNile. https://www.knowledgenile.com/blogs/the-rise-of-ai-meditation-apps-and-their-advantages

Kpmg and ping identity deliver consumer identity and access management solutions. (2018, January 2). News Release Archive. https://press.pingidentity.com/2018-01-02-KPMG-and-Ping-Identity-Deliver-Consumer-Identity-and-Access-Management-Solutions

Lee, M. (2023, August 31). *Ai personalization examples and challenges.* Bloomreach. https://www.bloomreach.com/en/blog/2023/ai-personalization-5-examples-business-challenges

Marr, B. (2023, January 13). The five biggest fitness and wellness technology trends in 2023. *Bernard Marr.* https://bernardmarr.com/the-5-biggest-fitness-and-wellness-technology-trends-in-2023/

Marr, B. (n.d.-a). *Artificial intelligence in space: The amazing ways machine learning is helping to unravel the mysteries of the universe.* Forbes. https://www.forbes.com/sites/bernardmarr/2023/04/10/artificial-intelligence-in-space-the-amazing-ways-machine-learning-is-helping-to-unravel-the-mysteries-of-the-universe/

Marr, B. (n.d.-b). *Debunking AI myths: The truth behind five common misconceptions.* Forbes. https://www.forbes.com/sites/bernardmarr/2023/07/05/debunking-ai-myths-the-truth-behind-5-common-misconceptions/

Marr, B. (n.d.-c). *The ten best examples of how AI is already used daily.* Forbes. https://www.forbes.com/sites/bernardmarr/2019/12/16/the-10-best-examples-of-how-ai-is-already-used-in-our-everyday-life/

Miller, K. (2023, January 23). *Designing ethical self-driving cars.* https://hai.stanford.edu/news/designing-ethical-self-driving-cars

Misas-Villamil, J. C., & Van Der Hoorn, R. A. (2008). Enzyme–inhibitor interactions at the plant–pathogen interface. *Current Opinion in Plant Biology, 11*(4), 380–388. https://doi.org/10.1016/j.pbi.2008.04.007

Mitigating bias in artificial intelligence. (n.d.). Berkeley Haas. https://haas.berkeley.edu/equity/resources/playbooks/mitigating-bias-in-ai/

Nielsen, J. (2023, July 16). *Ai improves employee productivity by 66%*. Nielsen Norman Group. https://www.nngroup.com/articles/ai-tools-productivity-gains/

Parsons, L. (2020, October 26). *Ethical concerns mount as AI takes a bigger role in decision-making*. Harvard Gazette. https://news.harvard.edu/gazette/story/2020/10/ethical-concerns-mount-as-ai-takes-bigger-decision-making-role/

Rado. (2024, May 15). 13 Best AI Project Management Software Tools for 2024. *Ayana*. https://ayanza.com/blog/ai-project-management-tools

Resource Library. (2024, March 22). https://www.powerschool.com/blog/ai-in-education/#:

Robinson, T. (2024, May 30). *Customizing nutrition plans: How AI can analyze your DNA and gut microbiome to craft your ideal diet*. https://www.linkedin.com/pulse/customizing-nutrition-plans-how-ai-can-analyze-your-dna-robinson--pag9c

Ryan-Mosley, T. (2023, February 20). *How AI can be helpful in disaster response*. MIT Technology Review. https://www.technologyreview.com/2023/02/20/1068824/ai-actually-helpful-disaster-response-turkey-syria-earthquake/

Sher, G., Benchlouch, A., Sher, G., & Benchlouch, A. (2023, October 31). The privacy paradox with AI. *Reuters*. https://www.reuters.com/legal/legalindustry/privacy-paradox-with-ai-2023-10-31/

Skipper, Dr. M. (2024, May 30). *Artificial Intelligence for Health: Opportunities, risks, and governance*. World Health Organization. https://www.who.int/news-room/events/detail/2024/05/30/default-calendar/artificial-intelligence-for-health-opportunities-risks-and-governance

Supercharging AI with quantum computing: A look into the future. (2023, December 16). Capitol Technology University. https://www.captechu.edu/blog/supercharging-ai-quantum-computing-look-future

The impact of AI technologies on the writing profession. (n.d.). *The Authors Guild*. https://authorsguild.org/advocacy/artificial-intelligence/impact/

The rise of AI-powered personal safety devices: Security's new frontier. (2023, May 4). Empowered by Ashley. https://us.empoweredbyashley.com/blogs/news/the-rise-of-ai-powered-personal-safety-devices-securitys-new-frontier

Ukonu, C. (2023, November 13). *4 ways AI can super-charge sustainable development*. World Economic Forum. https://www.weforum.org/agenda/2023/11/ai-sustainable-development/

Uppal, M. (2023, March 28). "*Networking: The secret weapon for career success in the AI era*." https://www.linkedin.com/pulse/networking-secret-weapon-career-success-ai-era-mohit-uppal

White, M. (2023, February 10). *13 AI Skills To Jumpstart Your AI Career in 2024*. Springboard. https://www.springboard.com/blog/data-science/ai-skills/

Winn, Z. (2023, November 29). They are *pushing the frontiers of art and technology with generative AI*. MIT News | Massachusetts Institute of Technology. https://

news.mit.edu/2023/pushing-frontiers-art-technology-generative-ai-1129

Yerushalmi, D. (n.d.). *Council post: How can ethics, regulations, and guidelines shape responsible AI?* Forbes. https://www.forbes.com/sites/forbestechcouncil/2024/02/05/how-ethics-regulations-and-guidelines-can-shape-responsible-ai/

www.ingramcontent.com/pod-product-compliance
Lightning Source LLC
Chambersburg PA
CBHW040924210326
41597CB00030B/5174